ANNALS *of* THE NEW YORK ACADEMY OF SCIENCES

T0179717

EDITOR-IN-CHIEF
Douglas Braaten

ASSOCIATE EDITOR
Rebecca E. Cooney

PROJECT MANAGER
Steven E. Bohall

EDITORIAL ADMINISTRATOR
Daniel J. Becker

Artwork and design by Ash Ayman Shairzay

The New York Academy of Sciences
7 World Trade Center
250 Greenwich Street, 40th Floor
New York, NY 10007-2157

annals@nyas.org
www.nyas.org/annals

**The New York
Academy of Sciences**

Published by Blackwell Publishing
On behalf of the New York Academy of Sciences

Boston, Massachusetts
2012

ANNALS *of* THE NEW YORK ACADEMY OF SCIENCES

VOLUME 1255

ISSUE

Annals Meeting Reports

TABLE OF CONTENTS

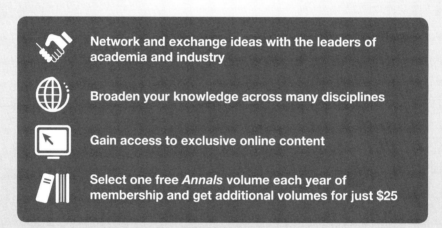

Ann. N.Y. Acad. Sci. ISSN 0077-8923

Diabetes and oral disease: implications for health professionals

David A. Albert,[1] Angela Ward,[1] Pamela Allweiss,[2,3] Dana T. Graves,[4] William C. Knowler,[5] Carol Kunzel,[1] Rudolph L. Leibel,[6] Karen F. Novak,[7] Thomas W. Oates,[8] Panos N. Papapanou,[1] Ann Marie Schmidt,[9] George W. Taylor,[10] Ira B. Lamster,[1] and Evanthia Lalla[1]

[1]Columbia University College of Dental Medicine, New York, New York. [2]Centers for Disease Control and Prevention, Atlanta, Georgia. [3]University of Kentucky College of Public Health, Lexington, Kentucky. [4]University of Pennsylvania School of Dental Medicine, Philadelphia, Pennsylvania. [5]National Institute of Diabetes and Digestive and Kidney Diseases, National Institutes of Health, Phoenix, Arizona. [6]Columbia University College of Physicians and Surgeons, New York, New York. [7]American Dental Education Association, Washington, D.C. [8]University of Texas Health Science Center at San Antonio, San Antonio, Texas. [9]New York University School of Medicine, New York, New York. [10]University of California at San Francisco School of Dentistry, San Francisco, California

Address for correspondence: David A. Albert DDS, MPH. daa1@columbia.edu Evanthia Lalla DDS, MS. el94@columbia.edu Columbia University College of Dental Medicine 630 West 168th Street, Box 20 New York, NY 10032

"Diabetes and Oral Disease: Implications for Health Professionals" was a one-day conference convened by the Columbia University College of Dental Medicine, the Columbia University College of Physicians and Surgeons, and the New York Academy of Sciences on May 4, 2011 in New York City. The program included an examination of the bidirectional relationship between oral disease and diabetes and the interprofessional working relationships for the care of people who have diabetes. The overall goal of the conference was to promote discussion among the healthcare professions who treat people with diabetes, encourage improved communication and collaboration among them, and, ultimately, improve patient management of the oral and overall effects of diabetes. Attracting over 150 members of the medical and dental professions from eight different countries, the conference included speakers from academia and government and was divided into four sessions. This report summarizes the scientific presentations of the event.[a]

Keywords: diabetes; oral disease; meeting report; complications; periodontal disease; management; interprofessional care

Introduction

A large portion of the U.S. population has periodontal disease, and this prevalence is significantly increased in individuals with diabetes. Evidence also suggests that diabetes leads to worsening periodontal disease, and, in turn, the systemic inflammation and infection that may result from periodontal disease can have an adverse effect on glycemic control and health outcomes, thus creating a cycle that compromises diabetes management in affected individuals. Any improvement in glycemic control and/or periodontal disease has the potential to make a significant impact on the quality of life for individuals with diabetes. In addition to periodontal disease, a number of different oral manifestations of diabetes have been documented.

Comprehensive diabetes care should be a team effort involving both the patient and a system of healthcare professionals. Improved communication between medical and dental care professionals can improve patient management of the oral and overall effects of the disease.

---

[a]The findings and conclusions in this report are those of the authors and do not necessarily represent the views of the organizations with whom the authors are affiliated.

doi: 10.1111/j.1749-6632.2011.06460.x
Ann. N.Y. Acad. Sci. 1255 (2012) 1–15 © 2012 New York Academy of Sciences.

Recently a symposium was held at the New York Academy of Sciences entitled "Diabetes and Oral Disease: Implications for Health Professionals," during which the bidirectional relationship between oral disease and diabetes was examined. The symposium's objective was to provide an opportunity for interactive and interdisciplinary discussion and education that would lead to enhanced quality of healthcare delivery, improved patient outcomes, and also serve as an impetus for medical and dental care professionals to coordinate and collaborate toward the goal of improving the health of individuals with diabetes.

Current concepts in diabetes

William C. Knowler, MD, DrPH (National Institute of Diabetes and Digestive and Kidney Diseases, National Institutes of Health), opened the symposium with a presentation entitled "The diabetes epidemic and the need for collaborative healthcare delivery."

To begin, Knowler spoke about the seriousness of diabetes as a chronic disease and explained that diabetes prevalence is dramatically increasing in most parts of the world. Morbidity and mortality, due to each major type of diabetes (type 1 and type 2), he continued, are primarily due to the long-term complications that have long been recognized to affect the eyes, kidneys, heart, blood vessels, and nerves. In addition, while long recognized as a complication of type 1 diabetes, periodontitis is also a complication of type 2 diabetes.[1,2] Periodontitis, Knowler maintained, is an important factor not only for oral health, but also for its association with many adverse health outcomes, presumably because it is accompanied by systemic inflammation. This seems to be supported by the findings from a longitudinal population study of Pima Indians ≥ 35 years old, in which diabetes and periodontal disease were assessed objectively. The researchers involved in this study reported that age- and sex-adjusted death rates from natural causes among diabetic persons were 3.7 deaths per 1000 person-years (95% confidence interval 0.7 to 6.6) in those with no or mild periodontal disease, 19.6 (10.7 to 28.5) in those with moderate periodontal disease, and 28.4 (22.3 to 34.6) in those with severe periodontal disease[3] (Fig. 1). This relationship remained significant when adjusted for numerous potentially confounding factors.

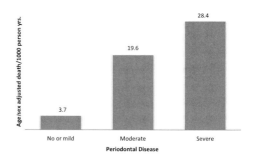

Figure 1. Mortality rates (all natural causes) in diabetic patients by periodontal disease status adjusted for age and sex to the 1985 Pima Indian population. Adapted from Saremi *et al.*[3]

Knowler continued by turning his attention to the prevention of diabetes. He described how the adverse consequences of periodontal disease in diabetes, as with all diabetes complications, would presumably be minimized or prevented if diabetes itself could be prevented. The prospects of preventing diabetes vary with the type of diabetes. The American Diabetes Association, he explained, classifies diabetes into two major types: 1 and 2.[4] Type 1 diabetes is due to autoimmune or idiopathic destruction of the pancreatic beta cells. Type 2 diabetes results from a combination of defects in insulin secretion and insulin action. Research in prevention of type 1 diabetes has focused on immune modulation, but so far has not met with reproducible success. By contrast, he continued, prevention research in type 2 diabetes, including lifestyle modification aimed at weight loss and increased physical activity or drugs affecting insulin resistance or secretion, has been at least partially successful.[5–7] Knowler concluded that because of the complex set of causes of diabetes, research in diabetes prevention requires a multidisciplinary scientific approach; and because diabetes affects many organ systems, preventing and treating diabetes complications requires collaboration among many healthcare professionals, including those in oral health.

Rudolph L. Leibel, MD (Naomi Berrie Diabetes Center, Columbia University College of Physicians and Surgeons), presented a talk entitled "Current concepts in the pathogenesis and treatment of diabetes." Leibel divided his presentation into two parts. First, he focused on type 2 diabetes, describing how the prevalence of type 2 diabetes is increasing rapidly in virtually all parts of the world, and that this increase almost perfectly parallels an increasing prevalence of obesity. Neither of these relatively acute changes in prevalence can be the result of much earlier genetic changes, but rather both reflect the consequences of genetic selection exercised on human populations. According to Leibel, as a species designed for environments in which high levels of physical activity and efficient storage of calories were required and in which lifespan was quite short, humans have clearly succeeded in creating man-made environments in which all of these predicates are eliminated. The consequences, Leibel continued, are pandemic obesity and type 2 diabetes.

The genes predisposing to fat accumulation and beta cell dysfunction appear to be largely distinct.[8,9] As a general formulation, the pathogenesis of type 2 diabetes involves interactions of multigenic susceptibilities of beta cells to metabolic stress. Obesity, by increasing resistance to insulin action and hence the need for insulin production by a limited number of beta cells, is perhaps the most important single stressor at present. The prevention of type 2 diabetes in individuals with impaired glucose tolerance, by exercise and weight reduction (NIH Diabetes Prevention Program, DPP),[5] and the reversal of recent-onset type 2 diabetes in obese individuals by moderate degrees of weight loss,[10] clearly demonstrates, Leibel explained, the functional relationship between obesity and type 2. Other environmental factors include intrauterine exposure to hyperglycemia and obesity, diet composition, and levels of physical fitness.

One of the major problems in applying a logical inference from these insights is that maintenance of a reduced body weight is extremely difficult for most patients. Part of the reason for this difficulty is the body's homeostatic mechanisms for the defense of body fat. These defenses are part of evolutionary biology and involve responses to reduced body weight that include reduced energy expenditure and increased drive to eat.[11] Full understanding of the physiology of these responses should lead to more effective prevention and treatment of obesity, and hence of type 2 diabetes.

In the second part of his presentation, Leibel spoke about type 1 diabetes—an autoimmune form of diabetes with a worldwide prevalence of about 10% of that of type 2 diabetes. And, like type 2 diabetes, the prevalence of type 1 diabetes is in-

creasing, though not as rapidly as type 2. Like type 2 diabetes, gene-by-environment interactions are dispositive. Most of the genetic susceptibility (and resistance) is conveyed by HLA genotypes that, in turn, mediate cellular immune responses to antigens that include insulin, but also environmental antigens not yet fully understood.[12,13] The natural history of type 1 diabetes includes a generally prolonged "run in" to the disease, as beta-cell mass is inexorably reduced by cellular immune assault on the beta cells of the islets of Langerhans. There is evidence that even in individuals with longstanding, insulin-requiring type 1 diabetes, residual beta cells are present in islets, resulting from ongoing beta-cell divisions that are simply not able to keep up with the persistent immune assault. The ability to enhance beta-cell replication to the point where the destructive process is outstripped could constitute an effective intervention. Likewise, suppression of the immune assault to the point where replication could compensate would be an alternative or synergistic approach. Finally, the advent of techniques to generate stem cells and to drive them toward beta-cell phenotypes could ultimately provide a source of beta cells that might be used to replace those subjected to immune destruction. These cells, if derived from a patient with type 1 diabetes, would likely still display the immune epitopes that are driving continued destruction of native beta cells. Steps would be required to silence the expression of these epitopes, or the presence of immunocytes directed at them, or both.

Our increased understanding of the genetic underpinnings of these two forms of diabetes, Leibel continued, enables us to "fix" these contributors in ways that enables clearer understanding of the environmental factors (and their mechanisms of action) that mediate the timing and severity of expression ("penetrance") of these genetic predispositions. Likewise, better understanding of the specifics of the environmental contributions permits better understanding of the mechanisms of genetic susceptibility. Leibel concluded by saying that science that exploits these reciprocal relationships will lead us to the insights required to prevent and cure these diseases.

The diabetes–oral disease connection

George W. Taylor, DMD, DrPH (University of California at San Francisco School of Dentistry), presented a paper on the bidirectional relationship between diabetes and periodontal disease. Taylor provided an epidemiologic perspective by reviewing the evidence for the adverse effects of diabetes on periodontal health, the role of periodontal infection in adversely affecting glycemic control, the impact of periodontal therapy on improving glycemic control, and the relationship of periodontal infection to the risk for developing diabetes complications, and possibly diabetes itself.

Taylor focused on some of the longitudinal observational studies that have provided evidence to support both the adverse effects of diabetes on periodontal health and those of severe periodontitis on increased risk for poorer glycemic control and diabetes complications.[14] He explained that the studies of the effects of nonsurgical periodontal therapy on glycemic control are a heterogeneous set of reports that include randomized clinical trials (RCTs) and clinical intervention studies that are not RCTs. Of the RCTs reported in the literature, several find a beneficial effect for periodontal therapy, although some RCTs did not. Recent meta-analyses of the intervention studies, Taylor explained, provided supporting evidence that nonsurgical periodontal therapy improves glycemic control, particularly in type 2 diabetes, with an average reduction of hemoglobin A1c of approximately 0.4% in pooled analyses[15,16] (Table 1). Taylor pointed out that this is a clinically important improvement because for each 1% reduction in mean HbA1c level, a 14–21% reduction in diabetes-related end points has been reported.[17]

In addition, Taylor spoke about the emerging evidence, from a small number of longitudinal observational studies, that suggests that periodontal disease is associated with increased risk for diabetes complications, including cardiovascular disease,[18] cardiorenal mortality,[3] and renal disease.[19] Taylor reported that there is evidence that periodontal infection may be a risk factor for the development of diabetes.[20]

One study, conducted by Saremi and colleagues,[3] followed a cohort of 628 Pima Indians in Arizona, USA, for a median follow-up time of 11 years. The researchers found those with severe periodontal disease at baseline had 3.2 times greater risk for cardiorenal mortality than those with no, mild, or moderate periodontal disease (Fig. 1). This estimate of significantly greater risk included controlling for several recognized major risk factors of cardiorenal mortality. A second study, Taylor explained, investigated the effect of periodontitis on risk for development of overt nephropathy (macroalbuminuria) and end-stage renal disease (ESRD) in a group of 529 Gila River Indian Community adults with type 2 diabetes. Shultis and colleagues[19] found that the incidence of macroalbuminura was 2.0, 2.1, and 2.6 times greater in individuals with moderate or severe periodontitis or in those who were edentulous, respectively, than those with no/mild periodontitis. The incidence of ESRD was also 2.3, 3.5, and 4.9 times greater for individuals with moderate or severe periodontitis or for those who were edentulous at baseline, respectively, than those with no/mild periodontitis.

Table 1. Effect of nonsurgical periodontal treatment on HbA1c levels in diabetes (meta-analysis of five studies, adapted from Teeuw *et al.*[15])[a]

Study	Tx Diff HbA1c B–E		C Diff HbA1c B–E		Weight %	WMD 95% CI
	N	Mean (SD)	N	Mean (SD)		
Katagiri 2009	32	−0.14 (0.63)	17	−0.09 (0.57)	27.42	−0.05[−0.40, 0.30]
Jones 2007	74	−0.65 (1.21)	80	−0.49 (1.22)	26.03	−0.16[−0.54, 0.22]
Kiran 2005	22	−0.86 (0.77)	22	0.31 (1.83)	12.77	−1.17[−2.00, −0.34]
Promsudthi 2005	27	−0.19 (0.74)	25	0.12 (1.05)	21.84	−0.31[−0.81, 0.19]
Stewart 2001	36	−1.90 (1.93)	36	−0.80 (1.85)	11.94	−1.10[−1.97, −0.23]
Total	191		180		100	−0.40[−0.77, −0.04]

[a]Weighted mean difference (WMD) of baseline (B) to end (E) in % HbA1c between treatment (Tx) and control (C) groups. Heterogeneity between 5 studies was 59.5%; test for heterogeneity: $Chi^2 = 9.87$, df = 4, $P = 0.04$. Test for overall effect: $Z = 2.15$, $P = 0.03$. Study references can be found in Ref. 15.

Demmer and colleagues[20] investigated the association of periodontal disease with the incidence of type 2 diabetes in over 7,000 participants of the First National Health and Nutrition Examination Survey (NHANES I) and the NHANES Epidemiologic Follow-up Survey. They reported a positive association between baseline periodontal disease and incident type 2 diabetes in a cohort study of individuals who were followed for a mean of 17 years. In addition, they found that periodontal disease was significantly associated with 50–100% greater risk for type 2 diabetes incidence at follow-up, after adjusting for other recognized risk factors for the development of type 2 diabetes.

In his conclusion, Taylor pointed out that the evidence, to date, supports the bidirectional, adverse relationship between periodontal infection and diabetes and that given the current evidence, it would be prudent to consider treating periodontal infection in people with diabetes as an important component of their overall management plan for diabetes care. However, Taylor emphasized that further rigorously conducted randomized clinical trials are necessary to unequivocally establish that treating periodontal infections can contribute to glycemic control and to the reduction of the burden of diabetes complications.

Ira B. Lamster, DDS, MMSc (Columbia University College of Dental Medicine), also briefly spoke about the association between periodontal disease and diabetes, but his presentation focused on the other oral and craniofacial disorders that have been associated with diabetes. Lamster explained that in addition to periodontal disease, dental caries, burning mouth syndrome, *Candida* infection, salivary dysfunction/xerostomia, taste and other neurosensory disorders, altered tooth eruption, and benign parotid hypertrophy all have been reported to be associated with diabetes.

The relationship of diabetes with oral health has an extensive literature that has been widely disseminated. Much of it, according to Lamster, has focused on periodontal disease, and comprehensive reviews have demonstrated that increased severity of periodontal disease is associated with diabetes mellitus.[14] Further, there is reasonable evidence to suggest that periodontitis is associated with poor metabolic control of diabetes, and that in the absence of other treatment, periodontal therapy can lead to a signif-

icant improvement in metabolic control (i.e., reduction in HbA1c) for a limited period of time (3 months).

Less attention, however, has been focused on the other oral complications of diabetes; yet it is essential that dental practitioners be aware of these disorders. Further, most of the clinical research has focused on young to middle-aged adults (25 to 55 years of age), with relatively limited research on younger individuals, or older adults.

Research from Columbia University on children and adolescents with diabetes (age range of 6 to 18 years), Lamster continued, has shown evidence of periodontal destruction. When compared to controls without diabetes, the relative risk of periodontal destruction was 2.72 (entire cohort, $P = 0.006$). When analyzed by age, younger individuals (ages 6–11) demonstrated greater risk (3.74, $P = 0.21$) than older individuals (ages 12–18; 2.63, $P = 0.066$).[21] Examining the relationship of diabetes-related parameters to the risk for periodontal destruction in this cohort of young patients with diabetes revealed that HbA1c was significantly associated with periodontitis, whereas duration of diabetes and BMI for age percentile were not.[22] Another report in this series of studies indicated that tooth eruption occurred sooner in young patients with diabetes as compared to nondiabetic controls. This occurred later in the eruption sequence (the extra-alveolar phase of tooth eruption).

With the aging of the population, the increased prevalence of diabetes in older adults, and the increased prevalence of oral diseases in the elderly, the study of oral manifestations of diabetes mellitus in older adults is both important and subject to confounds. Some of the more recent findings Lamster described included (a) that coronal caries were comparable in cases and controls, but the prevalence of root caries was higher in patients with diabetes—similarly, salivary flow was comparable in cases and controls, but the effects of xerogenic medications was more pronounced in patients with diabetes than controls;[23] (b) that for older edentulous patients with diabetes, a greater prevalence of burning mouth syndrome, dry mouth, angular cheilitis, and glossitis was observed, as compared to controls;[24] and (c) that benign parotid hypertrophy has been reported in older patients with diabetes. The prevalence is unknown, but preliminary evidence suggests that this is related to an enlargement of acinar cells, perhaps associated with an interruption in protein synthesis and release.

Lamster concluded his presentation by pointing out that a great deal has been learned about the oral complications of diabetes, but that much remains to be studied. The variety of oral lesions associated with diabetes emphasizes the importance of these disorders for patients, and for the dental professionals who care for them.

Karen Novak, DDS, MS, Ph.D. (American Dental Education Association), presented work on gestational diabetes mellitus and periodontitis and examined the possible link between these two conditions and maternal/fetal negative outcomes. Novak began by explaining that gestational diabetes mellitus (GDM) is a type of diabetes that develops during pregnancy and may or may not continue following parturition. Of the diabetes seen in pregnancy, only 10% is pregestational, with the remaining 90% being gestational.[25] GDM affects approximately 135,000 pregnant women (3–5%) annually in the United States, Novak explained, making it the most common metabolic disorder and medical complication of pregnancy.[26] Defined risk factors for development of GDM, she continued, include obesity, a family history of diabetes, having given birth previously to a very large infant, a still birth or a child with a birth defect, having too much amniotic fluid (polyhydramnios), and being older than 25 years of age.[27]

According to Novak, women with GDM make sufficient amounts of insulin; however, placental hormones (e.g., estrogen, cortisol, human placental lactogen) block the effect of insulin, leading to "insulin resistance." This begins about midway (20–24 weeks) through pregnancy. The larger the placenta grows, the more these hormones are produced, and the greater the insulin resistance becomes. In most women the pancreas is able to make additional insulin to overcome the insulin resistance, but when the pancreas makes all the insulin it can and there still is not enough to overcome the effect of the placenta's hormones, that is when GDM results, Novak explained.[28,29]

GDM can have a negative impact on both the mother and the fetus. Negative maternal outcomes associated with gestational diabetes include preeclampsia (hypertension), premature rupture of membranes, Caesarean section, and preterm delivery.[30–32] Although GDM develops or is discovered during pregnancy, and usually disappears when the pregnancy is over, 30–50% of women who have had GDM develop documented type 2 diabetes 3–5 years postpartum.[33,34] Moreover, greater acute and chronic neonatal morbidity and mortality have been described in neonates delivered by women with GDM.

Novak continued by saying that there is substantial evidence available documenting that the severity of periodontal disease is increased in patients with type 2 diabetes,[14,35] but minimal data are available on the effects of GDM on periodontal health. In addition, although substantial data have been accrued to support earlier observations that the infection and inflammation associated with periodontal disease may have a negative impact on the period of gestation and on fetal growth[36–38] there are limited data on the relationship among diabetes, periodontal disease, and pregnancy outcomes (combined effect). It was with this in mind, that Novak and her colleagues set out to study the hypothesis that women with GDM are at higher risk for developing more severe periodontal disease than women without GDM and that the combination of GDM and periodontal disease will be associated with an increased negative impact on maternal and fetal health.

Novak continued by describing the study in which women with GDM and non-GDM pregnant controls were recruited from the Division of Maternal-Fetal Medicine, Department of Obstetrics and Gynecology at the University of Kentucky. Subjects were matched based on age, gestational age, and race/ethnicity. Comprehensive medical and dental histories were obtained, and a periodontal examination, consisting of plaque index, probing pocket depths (PD), clinical attachment levels (LOA), bleeding index (BOP), and calculus index, was completed. Patients were further categorized as either having or not having periodontal disease. Periodontal disease was defined as having at least 4 teeth with PD $\geq$ 4 mm, LOA $\geq$ 2 mm, and BOP. Postdelivery maternal outcomes were evaluated as a composite, with the presence of any one of the following being a recorded negative outcome: preeclampsia, premature labor, premature rupture of membranes, urinary tract infections, chorioamnionitis/funisitis, induction of labor, operative vaginal deliveries, or unplanned cesarean.[39] Similarly, fetal outcomes were evaluated as a composite, with the presence of any one of the following constituting a negative outcome: intrauterine growth restriction/low-birth-weight, shoulder dystocia, brachial plexus damage, facial nerve injury, fractured bones, other neonatal birth problems, hypoglycemia, hyperbilirubinemia, respiratory distress syndrome, transient tachypnea of the newborn, polycythemia, hypocalcemia, intraventricular hemorrhage, necrotizing enterocolitis, congenital anomaly, stay in NICU, Apgar at 1 minute, or Apgar at 5 minutes of less than 7. Multiple logistic regression analyses, adjusted for smoking and calculus as known risk factors associated with periodontal disease are used to calculate odds ratios for adverse maternal and fetal outcomes. Women with periodontal disease and gestational diabetes serve as the reference group.

Thomas W. Oates, DMD, PhD (University of Texas Health Science Center at San Antonio), presented a paper on diabetes and its impact on dental implant therapy. Periodontal disease frequently results in tooth loss, with the cumulative effects most significant in older patients. It is in these older patients, who are particularly susceptible to type 2 diabetes and its comorbidities, that diabetes has been shown to significantly increase the levels of periodontal disease and tooth loss. Thus, one of the more subtle complications of diabetes may be a decrease in a patient's quality of life due to tooth loss and compromised masticatory function.[40]

Oates pointed out that oral health is an integral part of nutritional well-being and systemic health. Chronic diseases such as diabetes, he explained, have oral sequelae that may lead to compromises in oral function, and oral function may importantly modulate dietary interventions critical to the overall management of diabetes.[23] From a medical standpoint, there is no doubt that long-term good glycemic control is critical to the patient's minimizing diabetes related comorbidities. However, good glycemic control may be dependent upon proper masticatory function, Oates explained. With diabetes contributing to oral pathologies and tooth loss, tooth replacement, as can be provided with implant therapy, may be an important contributor to the patient's overall well-being. While diabetes remains a relative contraindication to implant therapy based on glycemic control, there are no strong clinical data supporting increased implant failures for patients lacking good glycemic control. In fact, more recent studies support the use of dental implant therapy for diabetic patients, even in the absence of good glycemic control.[41–44] Therefore, Oates explained, with the potential benefit implant therapy has to offer, it may be in the diabetes patient's best interest to consider implant therapy. While this represents a shift in attitudes toward diabetic patient care, it is, according to Oates, one that requires careful consideration of the risks and benefits of care, as well as the limitations in our understanding of this relationship.

Unraveling the mechanistic links between periodontitis and diabetes

Ann Marie Schmidt, MD (New York University School of Medicine), presented a paper on the role played by the receptor for advanced glycation endproducts (RAGE) and inflammation in diabetic complications, including periodontal disease.

RAGE is a multiligand receptor of the immunoglobulin superfamily. It was discovered as a receptor for advanced glycation endproducts (AGEs), the products of nonenzymatic glycation and oxidation of proteins and lipids that accumulate in diabetes. RAGE also binds proinflammatory ligands, such as members of the S100/calgranulin family, the high-mobility group box 1 (HMGB1), and amyloid-β peptide and β-sheet fibrils.[45] Strategies to block RAGE, such as soluble RAGE, the extracellular ligand-binding decoy of the receptor, or genetically modified mice, such as homozygous RAGE-null animals have been employed in various animal models of diabetes and its complications. These studies revealed that RAGE plays key roles in the development of macro- and microvascular complications in diabetes.[45] In this context, subjects with diabetes display increased severity of periodontal disease.

Schmidt continued by reporting findings of her and colleagues. They were able to show that RAGE and AGEs were expressed in human diabetic gingival tissue (retrieved at the time of periodontal surgery) and colocalized to both vascular and inflammatory cells.[46] Because similar findings were observed in diabetic mice inoculated by oral/anal gavage with the periodontal pathogen *Porphyromonas gingivalis* (Pg 381), Schmidt and colleagues tested the role of RAGE in periodontal disease in diabetic mice. They observed that, compared to nondiabetic mice, diabetic mice displayed significantly greater degrees of alveolar bone loss and gingival inflammation; in parallel, levels of gingival tissue inflammation and matrix metalloproteinases were higher in the diabetic tissues. Consistent with contributory roles for RAGE, administration of soluble RAGE suppressed exaggerated gingival inflammation, matrix metalloproteinase activity, and alveolar bone loss in *Pg* 381–infected mice.[47]

Schmidt concluded by reporting on a recent study by Lalla and colleagues. In this study the researchers addressed whether RAGE contributed directly to vascular inflammatory stress stimulated by *Pg* 381. During the study, endothelial cells were retrieved from the aortas of wild-type or RAGE-null mice and infected with *Pg* 381. When wild-type endothelial cells were infected with *Pg* 381, increased expression of RAGE, AGEs, and monocyte chemoattractant peptide-1 (MCP-1) resulted; however, in RAGE-null endothelial cells, *Pg* 381 infection did not elicit these findings.[48] Upregulation of these inflammatory mediators was also prevented by infection with DPG3, a fimbriae-deficient mutant of *Pg* 381, thereby suggesting that RAGE might contribute to invasion by this microorganism. Further experimentation, Schmidt conceded, is required to address this point. Taken together, these data link RAGE to *Pg*-dependent mechanisms that both destroy alveolar bone and stimulate endothelial cell stress—processes linking RAGE to the causes and consequences of periodontal inflammation and damage.

Dana T. Graves, DDS, MSc (University of Pennsylvania School of Dental Medicine), spoke about the impact of diabetes on inflammation, cell death, and bone in periodontal disease. Graves explained that diabetes mellitus is a metabolic disorder associated with several complications, including impaired healing. An important aspect of diabetes, and related to impaired healing is the increase in production of proinflammatory mediators, which include reactive oxygen species (ROS), advanced glycation endproducts, and cytokines such as TNF-α. Graves continued by explaining that the penetration of bacteria into connective tissue produces a significantly elevated inflammatory response in diabetic animals compared to nondiabetic controls. Microarray analysis further indicates that bacteria stimulate greater upregulation of a number of proinflammatory and proapoptotic genes in diabetic an-

imals, compared to normoglycemic controls. This, Graves explained, is due to a general dysregulation of cytokines upon bacterial perturbation, which can be reversed by inhibition of TNF-α[49]—a factor that has significant implications in both wound healing and periodontal disease. Soft tissue wounds of the skin and gingiva in diabetic animals are characterized by greater levels of inflammation, reduced proliferation, and greater apoptosis.[50] These aspects of diabetic wound healing can be reversed by inhibition of TNF-α, mechanistically linking reduced proliferation, greater apoptosis, and impaired healing to the effect of enhanced inflammation that is found in diabetic wounds.

Graves continued his presentation with an overview and examination of the impact diabetes has on periodontal disease. He explained that periodontal disease is significantly greater in individuals with diabetes and that it has been reported that diabetes increases the risk as well as the severity of periodontal disease. This position, Graves pointed out, is also evident in several animal models of diabetes. Diabetic animals exhibit enhanced bone loss and greater inflammation in experimental periodontitis.[47,51] In particular, diabetes appears to cause prolonged inflammation in various animal models following exposure to periodontal pathogens, suggesting difficulty in downregulating the inflammatory response. Graves reported that he and his colleagues investigated whether diabetes primarily affects periodontitis by enhancing bone loss or by limiting osseous repair using a ligature-induced model in the type 2 Zucker diabetic fatty (ZDF) rat and normoglycemic littermates.[51] They found that diabetes increased the intensity and duration of the inflammatory infiltrate. In addition, they found that while the formation of osteoclasts and bone resorption was initially similar in the diabetic animals, after the etiologic factor was removed, osteoclastogenesis persisted in the diabetic animals while it quickly returned to normal levels in the normoglycemic group. Moreover, Graves explained, the impact of diabetes on bone loss in periodontitis is further enhanced by interfering with coupled bone formation. Following an episode of periodontal bone loss, a certain amount of resorbed bone is regained by coupled bone formation. Bone coupling occurs, he explained, because bone is programmed to undergo a process of repair following bone loss, leading to a discrete level of regeneration. In experimental periodontitis, however, the amount of bone formation that occurs is incomplete and does not equal the amount of bone resorbed, leading to net bone loss.[49] When the amount of new bone formation following resorption was measured in diabetic and normoglycemic animals, the level was found to be 2.5-fold higher in the normal group ($P < 0.05$). Diabetes also increased apoptosis in bone-lining cells and periodontal ligament fibroblasts ($P < 0.05$). Thus, diabetes caused a more persistent inflammatory response, greater loss of attachment, more alveolar bone resorption, impaired coupled bone formation, and increased net bone loss. Graves concluded by reporting that recent studies indicate that diabetes can affect coupled bone formation by reducing proliferation in bone lining cells and by reducing the expression of growth factors that stimulate these cells, including basic fibroblast growth factor and transforming growth factor-beta.

Figure 2 presents a model of the potential mechanisms discussed by Schmidt and Graves in this session.

Interprofessional relationships in patient care

Evanthia Lalla, DDS, MS (Columbia University College of Dental Medicine), began her presentation by stating that in order to provide comprehensive care to people who have diabetes, there must be a team effort that involves the patient and various healthcare professionals. This group effort in the management of affected individuals is essential if more efficient and effective care is to be achieved.

The effort essentially begins with the patient, Lalla explained. The patient with diabetes needs to commit to self-care, make ongoing decisions regarding self-care and communicate frequently and honestly with healthcare providers. The healthcare professional's role in the team effort is to provide diabetic patients with guidance in goal setting, suggest strategies and techniques on how to achieve goals and overcome barriers, provide skills training (self-management techniques), screen, and manage the risk for complications.

Dental professionals in particular, Lalla continued, must discuss with their patients the link between oral and general health, how diabetes and periodontitis interrelate, and the need for comanagement of their condition by multiple healthcare providers, given that studies suggest that oral disease awareness among diabetic individuals is rather low.[52–57] They must also promote lifestyle changes and good oral and overall health behaviors. Special consideration regarding the treatment of dental patients with diabetes must also be taken into account in order to ensure that the oral care provided is safe and that therapeutic outcomes are predictable. These considerations include taking a thorough medical history, establishing

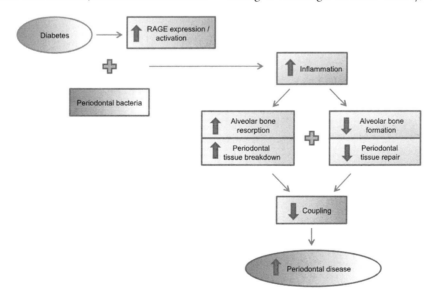

Figure 2. A model for the pathogenesis of enhanced periodontal disease in diabetes. Abbreviation: RAGE, receptor for advanced glycation endproducts.

communication with the treating physician, and performing a careful intraoral evaluation, including a comprehensive periodontal assessment. Initial therapy should focus on the control of acute infections, and a less complex, stepwise therapy plan should be offered when possible. Prevention, early recognition, and proper management of emergencies also are very important for dental professionals to address, Lalla pointed out. Dentists need to remember that for all people with type 1, and many with advanced type 2 diabetes, hypoglycemia is a fact of life. The hyperglycemic crisis is less common, but serious. Dental professionals, therefore, must consider timing and duration of appointments, possible need for change in diabetic regimen in consultation with the treating physician, and must provide profound anesthesia and pain control in conjunction with procedures and along with any antibiotic or host modulation agents. Clinical protocols and guidelines should be in place in every dental practice setting for determining frequency of follow-up care, determining the need for referral to a dental specialist, and for the need for medical consultation, referral, and follow-up.

Lalla continued by turning attention to the growing number of people in the United States who have diabetes but who remain undiagnosed. She explained that 70% of Americans see a dentist at least once per year[58] and that these patients often return for multiple, nonemergency, visits. This suggests that dental settings can also be healthcare locations actively involved in screening for unidentified diabetes. Dental professionals can assess risk factors, refer for testing or "formally" screen, and follow-up on outcomes. Borrell and colleagues[59] first explored the ability of clinical periodontal findings, coupled with self-reported information readily obtained during an individual's medical history review, to identify patients with undiagnosed diabetes (i.e., those with an FPG $\geq$ 126 mg/dl among those who responded negatively to the question, Have you ever been told by a doctor that you have diabetes?). Data from NHANES III public-use files (4,830 subjects $\geq$ 20 years of age) were used to calculate the predicted probability of having undiagnosed diabetes; findings suggested that such an approach was promising. This was subsequently corroborated by two other NHANES-based studies.[60,61] The first study to prospectively collect data in a clinical setting in order to single out a simple and efficient protocol

to identify people with undiagnosed pre-diabetes or diabetes, revealed that two dental parameters (number of missing teeth and percent of deep periodontal pockets) were effective in correctly identifying the majority of cases with unrecognized dysglycemia.[62] The addition of a point-of-care HbA1c test result was found to significantly improve the performance of the screening algorithm in the population under investigation.

As to the role of medical care providers, Lalla reported that the evidence to date suggests that physicians and nurses do not receive adequate training in oral health, are not comfortable performing a simple periodontal examination, and rarely advise patients on aspects of oral health.[63–67] Medical care providers need to discuss with their diabetic patients the importance of oral health and its relationship to diabetes and the potential sequelae of long-standing, untreated oral infections. All diabetic patients, she continued, should be advised to see a dentist on a regular basis. Screening for oral/periodontal changes must be part of the assessment of diabetic patients, similar to the screening for other complications. Asking about symptoms and performing a visual assessment of the mouth is simple and should be a part of the medical provider–patient interaction. In addition, medical care providers should also facilitate communication with the treating dentist by offering information on the patient's medical background, level of glycemic control, and presence of other complications and comorbidities; they should also be available to offer advice on medical management modifications that may be necessary.

Lalla concluded by saying that an interdisciplinary approach and collaboration beyond professional boundaries must become the standard of care for the management of the patients with, or at risk for, diabetes.

Carol Kunzel, PhD, MA (Columbia University College of Dental Medicine), presented a talk on her work, entitled "Dentists' attitudes and orientations in the management of the patient with diabetes." Kunzel began by explaining that because diabetes is a risk factor for periodontal disease, dentists can help reduce this risk by assessing, advising, and closely monitoring the diabetic patient. In doing so, dentists assume functions characteristic of primary and preventive healthcare clinicians. Thereby, the dental setting can be a healthcare location actively involved in identifying undiagnosed diabetic patients and assisting in the better management of diagnosed patients with diabetes.

In previous work, Kunzel and colleagues thought of this active dentists' involvement as having three phases: assessment, discussion, and active management.[68] Assessment, she explained, constitutes the dentists' asking the diabetic patient about the type and severity of disease; discussion represents their communication concerning diabetes with the patient; and active management reflects proactive actions taken to manage their diabetic status. Kunzel and colleagues investigated dentists' performance of these activities and their attitudes toward performing them, via a mail survey of representative samples of randomly selected dental general practitioners and periodontists in the northeast United States (GP response rate = 80%; periodontist response rate = 73%). Sample members were mailed a four-page questionnaire containing closed-ended items concerning attitudes and orientations regarding performing the three types of involvement, that is, assessment, discussion, and active management.[68]

Survey results indicated that general dentists are more willing to manage the care of diabetic patients on an assessing/advising basis than on a more active management basis.[68] With respect to periodontists, it was found that this pattern of active involvement with diabetic patients continued, although, overall, periodontists performed active management behaviors more frequently than general dentists.[69] Kunzel explained that when assessed internationally in a representative sample of general dentists in New Zealand, this pattern of involvement with the diabetic patient was again found. Most general dentists in New Zealand participated in the discussion phase of managing patients with diabetes, but the prevalence of involvement in active management activities was lower.[70]

Kunzel reported that, from an attitudinal perspective, the survey results showed that general dentists do not feel that they have mastery of the knowledge or behavioral areas involved; that viewing such activities as peripheral loomed as a barrier to performing them; and that they did not believe that their colleagues or patients expected them to perform such activities.[68] Like general dentists in New Zealand, approximately half of U.S. general dentists viewed more active management of patients

with diabetes as the responsibility of others. Also, more than half, like those in New Zealand, believed that taking an active role in diabetes management was useful, but only about a quarter (approximately 25%) thought it was easy. A minority in both the United States and New Zealand believed that their colleagues expected them to take a more active role in diabetes management.[70] Periodontists' attitudes were not clearly different than those of their general dentist colleagues.[69]

When dentists were asked about their willingness to perform certain active management activities the results varied depending on the activity involved. When dentists were asked about their willingness to screen for diabetes with a finger-stick test, relatively low levels of willingness were indicated. Kunzel suggested that these low levels may reflect concern over regulatory issues regarding the use of the test, or they may reflect dentists' reluctance to prick fingers to obtain a blood sample. Survey results also showed that more than 85% of periodontists expressed a strong willingness to refer a patient for such an evaluation, which, Kunzel proposed, suggests substantial inclination on their part to screen patients for undiagnosed diabetes, while a more moderate, but substantial, 69% of general dentists also expressed strong willingness do so.[70]

To further their understanding of how to encourage more involvement in active management, Kunzel and colleagues developed predictive models to identify explanators of general dentists' and periodontists' active management of the diabetic patient.[71] They found that general dentists were more influenced by the nature of their relationship with, and the characteristics of, their patients, while periodontists were more influenced by the nature of their relationship with their colleagues.

Kunzel concluded by posing a rhetorical question: Dare we be optimistic that the percentages of dentists, both general and specialist, who adopt more active management for the diabetic patient, will grow in the future? Her response showed some optimism. Kunzel pointed out that in New Zealand younger dentists seem to believe that their colleagues expect them to take a more active role in diabetes management.[70] This difference may reflect changes in the dental curriculum over time; perhaps there is a greater understanding of the general oral health connection present in the curriculum. As for the United States, Kunzel pointed out that there are indications that periodontists can play a leadership role in adoption efforts, since they, in higher percentages, are involved in active management of the diabetic patient.[69] Also suggested is the adoption of an incremental approach in which clinicians are first encouraged to become more actively engaged in discussion with the patient, because those who actively discuss tend to actively manage their diabetic patients.[71] It is hoped that such understandings, along with others, can contribute to diminishing a possible gap between the growth of science and the adoption of practice in this realm of patient care.

In her presentation titled "Working across medical–dental professional boundaries in the management of diabetes and its complications," Pamela Allweiss, MD, MPH (Centers for Disease Control and Prevention, and University of Kentucky College of Public Health), presented examples of how medical and dental professionals can work together to care for people with diabetes and how the resources in the public domain help this partnership.

Allweiss began by first giving the key points of consideration for the medical and dental professionals partnership. The points included (a) challenges in coordination of care when delivered by multiple providers in a variety of settings; (b) coordination to help ensure adherence to the intended treatment plan and identify drug and disease management problems in a timely manner; and (c) dental care professionals as primary points of care for people with diabetes.

Allweiss explained that the need for team care for people with diabetes that includes dental professionals and other healthcare professionals, such as pharmacists, podiatrists, and optometrists, is already being addressed by the National Diabetes Education Program (NDEP), a joint initiative of the Centers for Disease Control and Prevention (CDC) and the National Institutes of Health (NIH). NDEP, which partners under the umbrella of the PPOD (pharmacists, podiatrists, optometrists, and dentists) has over 200 public and private partners from multiple sectors (public health systems, and community programs, especially targeting populations with a large burden of diabetes) and is involved in the development and dissemination of evidence-based, focus group–tested materials that include diabetes control and prevention messages.

Two resources created by the NDEP include the PPOD Primer tool and the PPOD Checklist. The

Figure 3. The National Diabetes Education Program (a joint initiative of the Centers for Disease Control and Prevention and the National Institutes of Health) primer tool created to promote proper diabetes management by pharmacists, podiatrists, optometrists, and dentists.

PPOD Primer tool (Fig. 3) was developed to educate multiple providers to focus on comprehensive, interdisciplinary diabetes care. The tool includes sections that are specific to each discipline and are designed to provide a quick "crash course" on each specialty and its relationship to diabetes. Each section is written for the "other" providers to read and focuses on educating each provider about the role the other professions play in the diabetes care team. Emphasis is placed on the importance of conducting routine exams for complication prevention, recognizing danger signs, making recommendations regarding referrals, reinforcing among patients the need for self-exams, and of course, the importance of metabolic control. The goal of the PPOD Primer tool is to provide consistent messages across the disciplines and to encourage collaboration and a team approach in the caring for people with diabetes.

The PPOD Primer tool also has a patient education component. The tool informs patients that periodontal infection may make it difficult to control diabetes and, conversely, that poor metabolic control may increase susceptibility to infection. In addition, the tool explains that patients who have diabetes may be more likely to get periodontal infections, that the infection may take longer to heal, and that untreated infection may lead to loss of teeth. The PPOD Primer tool is widely disseminated using a variety of methods, some of which include distribution at professional meetings and continuing education programs, at professional associations, and on the NDEP website and the National Diabetes Information Clearinghouse.

The PPOD Checklist, Allweiss continued, is a tool developed by PPOD providers and other healthcare professionals, such as physicians, physicians' assistants, and nurses. Its goals are to ease communication among multiple providers, educate people with diabetes about needed exams, and help to improve pay-for-performance measures. During the development of the checklist, the PPOD working group conducted a pilot test of a Multidisciplinary Patient Care Checklist. Individual working group members sent the checklist to coworkers and colleagues and invited them to comment through a SurveyMonkey questionnaire. The goal of the pilot is to gauge whether the checklist would be useful, and used, in a real-world clinical setting. Most respondents agreed that the content was appropriate and presented clearly. In addition, 74.3% responded that they were likely to change their practice to more of a team approach, incorporating the members of the team, or by referral. In closing, Allweiss reported that the survey responses revealed that the checklist is useful in actual practice, with many (30%) indicating its potential application in EMR/EHR systems. It is currently being pilot tested in an electronic medical records format and will eventually be available on the NDEP website (www.yourdiabetesinfo.org).

Conclusion

The 2011 Columbia University/New York Academy of Sciences conference on diabetes and oral disease brought together clinicians and researchers from medicine and dentistry, and provided a setting for education and interaction, aiming to increase awareness and collaboration across disciplines. Conference speakers covered information on the demographics, epidemiology, pathophysiology, and treatment of diabetes and periodontitis. They explained the factors that constitute the bidirectional diabetes–oral disease link and defined the role of oral disease in initiating the inflammatory response, as well as the impact of hyperglycemia on oral health. In addition, speakers presented information on how to screen and counsel patients for oral disease and diabetes risk and emphasized that interprofessional patient management is essential in order to achieve improved health outcomes in affected individuals.

Acknowledgment

Dr. Knowler is supported by the Intramural Research Program of the National Institute of Diabetes and Digestive and Kidney Diseases.

The "Diabetes and Oral Disease: Implications for Health Professionals" conference was made possible by generous support from The New York Academy of Sciences (NYAS) Silver Supporter, Oral Health America; NYAS Bronze Supporters, Aetna Dental® and Colgate-Palmolive Co.; and NYAS Academy Friend, the Institute for Oral Health. The event was also funded in part by the Life Technologies™ Foundation.

Open access of the meeting report is provided by The Sackler Institute for Nutrition Science at The New York Academy of Sciences. [Correction added on 04/17/2012 after initial online publication.]

Conflicts of interest

The authors declare no conflicts of interest.

References

1. Shlossman, M. *et al.* 1990. Type 2 diabetes mellitus and periodontal disease. *J. Am. Dent. Assoc.* **121:** 532–536.
2. Nelson, R.G. *et al.* 1990. Periodontal disease and NIDDM in Pima Indians. *Diabetes Care* **13:** 836–840.
3. Saremi, A. *et al.* 2005. Periodontal disease and mortality in type 2 diabetes. *Diabetes Care* **28:** 27–32.
4. American Diabetes Association. 2011. Diagnosis and classification of diabetes mellitus. *Diabetes Care* **34:** 62–69.
5. Knowler, W.C. *et al.* 2002. Reduction in the incidence of type 2 diabetes with lifestyle intervention or metformin. *N. Engl. J. Med.* **346:** 393–403.
6. Crandall, J.P. *et al.* 2008. The prevention of type 2 diabetes. *Nature Clin. Pract. Endocrinol. Metab.* **4:** 382–393.
7. DeFronzo, R.A. *et al.* 2011. Pioglitazone for diabetes prevention in impaired glucose tolerance. *N. Engl. J. Med.* **364:** 1104–1115.
8. Li, S. *et al.* 2010. Cumulative effects and predictive value of common obesity-susceptibility variants identified by genome-wide association studies. *Am. J. Clin. Nutr.* **91:** 184–190.

9. Florez, J.C. 2008. Newly identified loci highlight beta cell dysfunction as a key cause of type 2 diabetes: where are the insulin resistance genes? *Diabetologia* **51:** 1100–1110.

10. Dixon, J.B. *et al.* 2008. Adjustable gastric banding and conventional therapy for type 2 diabetes: a randomized controlled trial. *J. Am. Dent. Assoc.* **299:** 316–323.

11. Leibel, R.L. 2008. Molecular physiology of weight regulation in mice and humans. *Int. J. Obesity (Lond.)* **32:** 98–9108.

12. Ermann, J. & C.G. Fathman. 2001. Autoimmune diseases: genes, bugs and failed regulation. *Nature Immunol.* **2:** 759–761.

13. Eisenbarth, G.S. 2010. Banting Lecture 2009: An unfinished journey: molecular pathogenesis to prevention of type 1A diabetes. *Diabetes* **59:** 759–774.

14. Taylor, G.W. & W.S. Borgnakke. 2008. Periodontal disease: associations with diabetes, glycemic control and complications. *Oral Dis.* **14:** 191–203.

15. Teeuw, W.J., V.E.A. Gerdes & B.G. Loos. 2010. Effect of periodontal treatment on glycemic control of diabetic patients: a systematic review and meta-analysis. *Diabetes Care* **33:** 421–427.

16. Simpson, T.C. *et al.* 2010. Treatment of periodontal disease for glycaemic control in people with diabetes. *Cochrane Database Syst. Rev.* **5:** CD004714.

17. Stratton, I.M. *et al.* 2000. Association of glycaemia with macrovascular and microvascular complications of type 2 diabetes (UKPDS 35): prospective observational study. *Br. Med. J.* **321:** 405–412.

18. Thorstensson, H., J. Kuylenstierna & A. Hugoson. 1996. Medical status and complications in relation to periodontal disease experience in insulin-dependent diabetics. *J. Clin. Periodontol.* **23:** 194–202.

19. Shultis, W.A. *et al.* 2007. Effect of periodontitis on overt nephropathy and end-stage renal disease in type 2 diabetes. *Diabetes Care* **30:** 306–311.

20. Demmer, R.T., D.R. Jacobs & M. Desvarieux. 2008. Periodontal disease and incident type 2 diabetes: results from the First National Health and Nutrition Examination Survey and its epidemiologic follow-up study. *Diabetes Care* **31:** 1373–1379.

21. Lalla, E. *et al.* 2007. Diabetes mellitus promotes periodontal destruction in children. *J. Clin. Periodontol.* **34:** 294–298.

22. Lalla, E. *et al.* 2007. Diabetes-related parameters and periodontal conditions in children. *J. Periodontal. Res.* **42:** 345–349.

23. Meurman, J.H. *et al.* 1998. Saliva in non-insulin dependent diabetic patients and control subjects. *Oral Surg. Oral Med. Oral Pathol. Oral Radiol. Endod.* **86**(1): 69–76.

24. Dorocka-Bobkowska, B. *et al.* 2010. Candida-associated denture stomatitis in type 2 diabetes mellitus. *Diabetes Res. Clin. Pract.* **90:** 81–86.

25. El-Sayed, Y.Y. & D.J. Lyell. 2001. New therapies for the pregnant patient with diabetes. *Diabetes Technol. Ther.* **3:** 635–640.

26. Coustan, D.R. 1995. Gestational diabetes. In *Diabetes in America*, 2nd ed. H. MI, *et al.*, Eds.: 703–771.

27. Kjos, S.L. & T.A. Buchanan. 1999. Gestational diabetes mellitus. *N. Engl. J. Med.* **341:** 1749–1756.

28. Samaan, N. *et al.* 1968. Metabolic effects of placental lactogen (HPL) in man. *J. Clin. Endocrinol. Metab.* **28:** 485–491.

29. Kalkhoff, R.K., M. Jacobson & D. Lemper. 1970. Progesterone, pregnancy and the augmented plasma insulin response. *J. Clin. Endocrinol. Metab.* **31:** 24–28.

30. Jensen, D.M. *et al.* 2000. Maternal and perinatal outcomes in 143 Danish women with gestational diabetes mellitus and 143 controls with a similar risk profile. *Diabet. Med.* **17:** 281–286.

31. Xiong, X. *et al.* 2001. Gestational diabetes mellitus: prevalence, risk factors, maternal and infant outcomes. *Int. J. Gynaecol. Obstet.* **75:** 221–228.

32. Sendag, F. *et al.* 2001. Maternal and perinatal outcomes in women with gestational diabetes mellitus as compared to nondiabetic controls. *J. Reprod. Med.* **46:** 1057–1062.

33. Mestman, J.H., G.V. Anderson & V. Guadalupe. 1972. Follow-up study of 360 subjects with abnormal carbohydrate metabolism during pregnancy. *Obstet Gynecol.* **39:** 421–425.

34. Metzger, B.E. & D.R. Coustan. 1998. Summary and recommendations of the Fourth International Workshop-Conference on Gestational Diabetes Mellitus. The Organizing Committee. *Diabetes Care* **21:** 161–167.

35. Chavarry, N.G.M. *et al.* 2009. The relationship between diabetes mellitus and destructive periodontal disease: a meta-analysis. *Oral Health Prev. Dent.* **7:** 107–127.

36. Khader, Y.S. & Q. Ta'ani. 2005. Periodontal diseases and the risk of preterm birth and low birth weight: a meta-analysis. *J. Periodontol.* **76:** 161–165.

37. Xiong, X. *et al.* 2006. Periodontal disease and adverse pregnancy outcomes: a systematic review. *BJOG* **113:** 135–143.

38. Vergnes, J.-N. & M. Sixou. 2007. Preterm low birth weight and maternal periodontal status: a meta-analysis. *Am. J. Obstet. Gynecol.* **196:** 1–7.

39. Crowther, C.A. *et al.* 2005. Effect of treatment of gestational diabetes mellitus on pregnancy outcomes. *N. Engl. J. Med.* **352:** 2477–2486.

40. McGrath, C. & R. Bedi. 2001. Can dentures improve the quality of life of those who have experienced considerable tooth loss? *J. Dent.* **29:** 243–246.

41. Dowell, S., T.W. Oates & M. Robinson. 2007. Implant success in people with type 2 diabetes mellitus with varying glycemic control: a pilot study. *J. Am. Dent. Assoc.* **138:** 355–361.

42. Tawil, G. *et al.* 2008. Conventional and advanced implant treatment in the type II diabetic patient: surgical protocol and long-term clinical results. *Int. J. Oral. Maxillofac. Implants.* **23:** 744–752.

43. Oates, T.W. *et al.* 2009. Glycemic control and implant stabilization in type 2 diabetes mellitus. *J. Dent. Res.* **88:** 367–371.

44. Turkyilmaz, I. 2010. One-year clinical outcome of dental implants placed in patients with type 2 diabetes mellitus: a case series. *Implant. Dent.* **19:** 323–329.

45. Ramasamy, R., S.F. Yan & A.M. Schmidt. 2009. RAGE: therapeutic target and biomarker of the inflammatory response—the evidence mounts. *J. Leukoc. Biol.* **86:** 505–512.

46. Schmidt, A.M. *et al.* 1996. Advanced glycation endproducts (AGEs) induce oxidant stress in the gingiva: a potential

mechanism underlying accelerated periodontal disease associated with diabetes. *J. Periodontal. Res.* **31:** 508–515.

47. Lalla, E. *et al.* 2000. Blockade of RAGE suppresses periodontitis-associated bone loss in diabetic mice. *J. Clin. Invest.* **105:** 1117–1124.

48. Pollreisz, A. *et al.* 2010. Receptor for advanced glycation end-products mediates pro-atherogenic responses to periodontal infection in vascular endothelial cells. *Atherosclerosis* **212:** 451–456.

49. Graves, D.T., J. Li & D.L. Cochran. 2011. Inflammation and uncoupling as mechanisms of periodontal bone loss. *J. Dent. Res.* **90:** 143–153.

50. Siqueira, M.F. *et al.* 2010. Impaired wound healing in mouse models of diabetes is mediated by TNF-alpha dysregulation and associated with enhanced activation of forkhead box O1 (FOXO1). *Diabetologia* **53:** 378–388.

51. Liu, R. *et al.* 2006. Diabetes enhances periodontal bone loss through enhanced resorption and diminished bone formation. *J. Dent. Res.* **85:** 510–514.

52. Allen, E.M. *et al.* 2008. Attitudes, awareness and oral health-related quality of life in patients with diabetes. *J. Oral Rehabil.* **35:** 218–223.

53. Tomar, S.L. & A. Lester. 2000. Dental and other health care visits among U.S. adults with diabetes. *Diabetes Care* **23:** 1505–1510.

54. Sandberg, G.E., H.E. Sundberg & K.F. Wikblad. 2001. A controlled study of oral self-care and self-perceived oral health in type 2 diabetic patients. *Acta Odontol. Scand.* **59:** 28–33.

55. Moore, P.A. *et al.* 2000. Diabetes and oral health promotion: a survey of disease prevention behaviors. *J. Am. Dent. Assoc.* **131:** 1333–1341.

56. Jansson, H. *et al.* 2006. Type 2 diabetes and risk for periodontal disease: a role for dental health awareness. *J. Clin. Periodontol.* **33:** 408–414.

57. Al Habashneh, R. *et al.* 2010. Knowledge and awareness about diabetes and periodontal health among Jordanians. *J. Diabetes Complications* **24:** 409–414.

58. Centers for Disease Control and Prevention. 2011. *National diabetes fact sheet: national estimates and general information on diabetes and prediabetes in the United States, 2010.* U.S. Department of Health and Human Services, Centers for Disease Control and Prevention. Atlanta, GA.

59. Borrell, L.N. *et al.* 2007. Diabetes in the dental office: using NHANES III to estimate the probability of undiagnosed disease. *J. Periodontal Res.* **42:** 559–565.

60. Strauss, S.M. *et al.* 2010. The dental office visit as a potential opportunity for diabetes screening: an analysis using NHANES 2003–2004 data. *J. Public Health Dent.* **70:** 156–162.

61. Li, S., P.L. Williams & C.W. Douglass. 2011. Development of a clinical guideline to predict undiagnosed diabetes in dental patients. *J. Am. Dent. Assoc.* **142:** 28–37.

62. Lalla, E. *et al.* 2011. Identification of unrecognized diabetes and pre-diabetes in a dental setting. *J. Dent. Res.* **90:** 855–860.

63. Quijano, A. *et al.* 2010. Knowledge and orientations of internal medicine trainees toward periodontal disease. *J. Periodontol.* **81:** 359–363.

64. Al-Habashneh, R. *et al.* 2010. Diabetes and oral health: doctors' knowledge, perception and practices. *J. Eval. Clin. Pract.* **16:** 976–980.

65. Al-Khabbaz, A.K., K.F. Al-Shammari & N.A. Al-Saleh. 2011. Knowledge about the association between periodontal diseases and diabetes mellitus: contrasting dentists and physicians. *J. Periodontol.* **82:** 360–366.

66. Yuen, H.K. *et al.* 2009. Oral health knowledge and behavior among adults with diabetes. *Diabetes Res. Clin. Pract.* **86:** 239–246.

67. Ward, A.S. *et al.* 2010. Application of the theory of planned behavior to nurse practitioners' understanding of the periodontal disease-systemic link. *J. Periodontol.* **81:** 1805–1813.

68. Kunzel, C. *et al.* 2005. On the primary care frontlines: the role of the general practitioner in smoking-cessation activities and diabetes management. *J. Am. Dent. Assoc.* **136:** 1144–1153.

69. Kunzel, C., E. Lalla & I.B. Lamster. 2006. Management of the patient who smokes and the diabetic patient in the dental office. *J. Periodontol.* **77:** 331–340.

70. Forbes, K. *et al.* 2008. Management of patients with diabetes by general dentists in New Zealand. *J. Periodontol.* **79:** 1401–1408.

71. Kunzel, C., E. Lalla & I.B. Lamster. 2007. Dentists' management of the diabetic patient: contrasting generalists and specialists. *Am. J. Public Health* **97:** 725–730.

Ann. N.Y. Acad. Sci. ISSN 0077-8923

ANNALS OF THE NEW YORK ACADEMY OF SCIENCES
Issue: Annals *Meeting Reports*

The Sixth Annual Translational Stem Cell Research Conference of the New York Stem Cell Foundation

Caroline Marshall,[1] Haiqing Hua,[1,2] Linshan Shang,[1] Bi-Sen Ding,[3,1] Giovanni Zito,[4,1] Giuseppe Maria de Peppo,[1] George Kai Wang,[2,1] Panagiotis Douvaras,[1] Andrew A. Sproul,[1] Daniel Paull,[1] Valentina Fossati,[1] Michael W. Nestor,[1] David McKeon,[1] Kristin A. Smith,[1] and Susan L. Solomon[1]

[1]The New York Stem Cell Foundation, New York, New York. [2]Columbia University, New York, New York. [3]Weill Cornell Medical College, New York, New York. [4]Yale University, New Haven, Connecticut

Address for correspondence: Caroline Marshall, The New York Stem Cell Foundation, 163 Amsterdam Avenue, Box 309, New York, NY 10023

The New York Stem Cell Foundation's "Sixth Annual Translational Stem Cell Research Conference" convened on October 11–12, 2011 at the Rockefeller University in New York City. Over 450 scientists, patient advocates, and stem cell research supporters from 14 countries registered for the conference. In addition to poster and platform presentations, the conference featured panels entitled "Road to the Clinic" and "The Future of Regenerative Medicine."

Introduction

The New York Stem Cell Foundation's (NYSCF) annual translational stem cell conference opened with two panels of experts discussing current progress in translating stem cell research to accelerate cures for the major diseases of our time. In the first panel "Road to the Clinic," moderated by Lee Rubin, director of translational medicine at the Harvard Stem Cell Institute and NYSCF scientific advisor, leading experts from the pharmaceutical, biotechnology, and healthcare industries were joined by representatives from venture capital and grant-awarding foundations to discuss how to transfer stem cell research from the laboratory into safe and effective treatments. Panelists included Stephen Chang (New York Stem Cell Foundation), Scott Johnson (Myelin Repair Foundation), Robert J. Palay (Cellular Dynamics International), and William A. Sahlman (Harvard Business School). The panel discussed the latest research developments and the scientifically challenging path toward clinical trials. The second panel, "The Future of Regenerative Medicine," composed of leading stem cell researchers and policy makers, discussed from their differing perspectives the regulatory challenges facing researchers on the road to the clinic. Moderator and chief executive officer of the New York Stem Cell Foundation Susan L. Solomon, and panelists Mahendra S. Rao (NIH Intramural Center for Regenerative Medicine), Craig B. Thompson (Memorial Sloan-Kettering Cancer Center), Marc Tessier-Lavigne (The Rockefeller University), and Irving L. Weissman (Stanford University) examined the changes in policy that will be needed to advance the development, evaluation, and approval of emerging stem cell treatments and regenerative medicine.

As in previous years, the second day of the conference convened an international panel of researchers at the forefront of the stem cell field who presented their work on diabetes, heart and muscles, cancer and blood disease, neurodegeneration, and programming/reprogramming.[1,2] The panel included a groundbreaking report on somatic cell nuclear transfer (SCNT), recently published in *Nature* by Dieter Egli at the NYSCF laboratory.

In addition to a keynote address by Irving L. Weissman, who discussed the latest developments in cancer treatments using stem cells, the inaugural recipient of the NYSCF–Robertson Stem Cell Prize, Professor Peter J. Coffey (University College London, and the London Project to

doi: 10.1111/j.1749-6632.2012.06481.x

Ann. N.Y. Acad. Sci. 1255 (2012) 16–29 © 2012 New York Academy of Sciences.

Cure Blindness) presented his forthcoming clinical trial using stem cells to treat age-related macular degeneration.

NYSCF, in collaboration with *Annals of New York Academy of Sciences*, is delighted to present this report; compiled by young stem cell scientists, it summarizes the excellent, groundbreaking progress in stem cell research.

Diabetes

The opening speaker, Hans Snoeck (Mount Sinai School of Medicine), discussed the importance of using stem cells to generate thymus tissue. *In vivo*, the thymus is the immune organ in which T cell positive and negative selection takes place. Thymus tissue derived *in vitro* has potential therapeutic applications, such as enhancing T cell reconstitution after allogeneic bone marrow transplantation. Humanized mouse models can be created by transplanting thymus generated from human stem cells, which are extremely valuable for studying human autoimmune diseases, including type 1 diabetes.[3] Snoeck showed that when using activin A, BMP4, and basic fibroblast growth factor (bFGF), his group differentiated human embryonic stem cells into anterior endoderm, from which thymic tissue is derived. Subsequent treatment with Noggin and SB-431542 directs endoderm cells toward anterior foregut fate, with more than 90% of the cells expressing SOX2.[4] By applying knowledge- and screening-based approaches, Snoeck's group discovered combinations of various factors, including Wnt3A, fibroblast growth factor 10 (FGF10), keratinocyte growth factor (KGF), and sonic hedgehog (SHH), that could further specify foregut endoderm cells to become lung bud or pharyngeal endoderm cells. Notably, Snoeck pointed out that Wnt3A is crucial to prepattern the cells for lung development, and that retinoic acid induces lung fate partly through suppression of TBX1, a transcription factor necessary for pharyngeal development.

Shuibing Chen (Weill Cornell Medical College; Fig. 1) summarized previous and ongoing efforts to generate pancreatic β cells from stem cells, including hypothesis-driven approaches based on animal models and *in vitro* studies that work well for the early stages of differentiation, specifically during the definitive endoderm and gut-tube endoderm stage.[5] In order to efficiently direct the cells into pancreatic progenitors and β cells, however,

she and her colleagues have been using discovery-driven screen approaches to search for the optimal conditions to facilitate this transition. After screening with chemical libraries, they identified a small molecule, (–)-indolactam V, that can promote the generation of pancreatic progenitors by activating protein kinase C (PKC) signaling.[6] Recently, Chen's group screened for growth factors and cell lines that induce pancreatic progenitor and β cell generation and showed that mouse pancreatic endothelial cells and a human endothelial cell line (AKT-HBVEC) significantly enhance the proliferation of pancreatic progenitors in culture. Furthermore, Chen found that cell–cell contact is not required and the effect is mediated by factors that belong to the epidermal growth factor (EGF) family. Currently, the insulin-producing β cells derived *in vitro* are immature β cells, which do not display great response to glucose. Chen's group is now seeking a cellular niche that can accelerate the proliferation and/or maturation of β cells.

The final talk in this session was given by Pedro Herrera (University of Geneva) who introduced the intriguing phenomenon of α cell to β cell transition. Herrera's group has developed new transgenic models allowing near-total α cell or β cell removal, specifically in adult mice; these mouse models enable the study of pancreas regeneration after massive cell loss. Herrera and colleagues showed that six months after β cell removal, new β cells were regenerated. Around 20% of newly generated β cells express both glucagon and insulin, hormones produced by α cells and β cells, respectively. Using cell-tracing technology, Herrera and colleagues demonstrated that some of the newly formed β cells were derived from α cells.[7] In contrast, they also observed that when most of α cells are removed, the remaining 2% of the normal α cell mass is enough to maintain healthy and euglycemic mice.[8] Taken together, these studies suggest that future diabetic therapies could involve regenerating β cells via reprogramming adult α cells.

Cancer and blood disease

Shahin Rafii (Howard Hughes Institute and Weill Cornell Medical College) opened this session and described a recently identified instructive role of pulmonary capillary endothelial cells (PCECs) in supporting lung regeneration.[9] Previous work from the Rafii group established the novel concept that

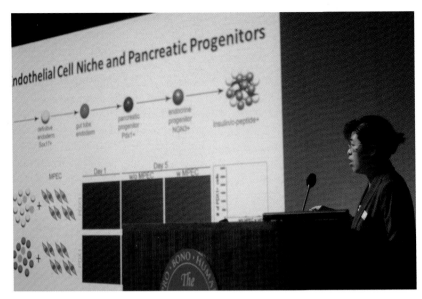

Figure 1. Shuibing Chen (Weill Cornell Medical College and NYSCF-Robertson investigator) presents her work on deriving functional pancreatic endocrine cells from human pluripotent stem cells.

capillary endothelial cells (CECs) not only function as passive conduits to meet metabolic demand, they also stimulate paracrine (angiocrine) growth factors to induce and sustain organ regeneration.[10–14] In bone marrow (BM)[10–12] and liver,[13] sinusoidal endothelial cells (SECs) constitute phenotypically and functionally distinct populations of cells. After partial hepatectomy, liver SECs—via angiocrine production of hepatocyte growth factor and Wnt2 (a process defined as *inductive angiogenesis*)—stimulate hepatocyte proliferation.[13] Subsequently, liver SECs undergo *proliferative angiogenesis* to match the increasing demand for blood supply in the regenerating liver. Likewise, after chemotherapy and irradiation, activated BM SECs induce hematopoiesis by angiocrine generation of Notch ligands and insulin-like growth factor binding proteins (IGFBPs).[10,12]

To investigate the role of PCECs in supporting lung regeneration, Rafii's group generated a unilateral pneumonectomy (PNX) mouse model by performing surgical resection of the left lung lobe. Without perturbing the vascular integrity of the remaining lobes, PNX induces significant growth of mass, volume, and physiological respiratory capacity of the remaining lungs. This regeneration process is due to neoalveologenesis, a process involving amplification of alveolar epithelial progenitor cells and PCECs. The phenotypic and op-

erational markers of mouse PCECs were defined as VE-cadherin+ VEGFR2+ FGFR1+ CD34+ ECs. They further demonstrated that PNX, through activation of VEGFR2 and FGFR1, induces PCECs of the remaining lobes to produce the angiocrine matrix metalloprotease MMP14. In turn, MMP14 promotes regenerative alveolarization by unmasking cryptic EGF-like ligands that stimulate proliferation of epithelial progenitor cells. These studies underscore the instrumental role of endothelial-derived angiocrine signals in instructing adult organ regeneration. Selective activation of CECs, or increasing the generation of angiocrine factors in patients, would promote organ regeneration, thereby offering therapeutic avenues for end-stage liver, lung, and hematopoietic diseases.

The second speaker of the session was Viviane Tabar (Memorial Sloan-Kettering Cancer Center), whose laboratory focuses on understanding the alterations within the brain stem cell niche that are responsible for glioblastoma tumors. Previously, it was demonstrated that a subpopulation of CD133+ cancer stem cells are responsible for the initiation of glioblastoma tumors.[15] Tabar's group discovered that CD133+ cancer stem cells not only initiate the tumor but also generate new endothelial cells necessary for tumor self-maintenance.[16] The newly generated cells presented a different genotype when compared with normal blood vessel

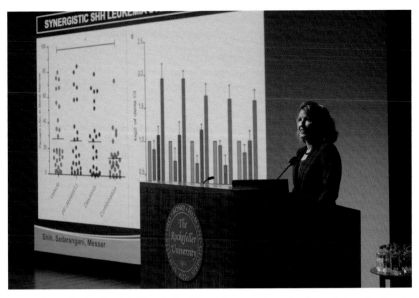

Figure 2. Catriona Jamieson (University of California, San Diego) discusses the molecular evolution of leukemia stem cells during the session entitled "Cancer and Blood Disease."

cells in the brain: the cells have an amplification of the EGF receptor gene and the centromere of chromosome 7, two typical mutations of different glioblastomas variants.[17] Tabar and colleagues isolated CD133[+] cells from human glioblastoma samples and cultured them *in vitro* to test their ability to generate tumor vessels. Interestingly, they found that after seven days *in vitro*, CD133[+] cancer stem cells differentiate into blood vessel cells expressing CD105, CD31, and CD34, which are typical endothelial markers in the brain. Tabar's group confirmed this *in vivo* by inoculating NOD/SCID mice with CD133[+] GFP[+] cells isolated from glioblastoma tumors. Their results showed that gliobastomas that developed in NOD/SCID mice were characterized by GFP[+] blood vessels cells, suggesting that the CD133[+] cancer stem cells were able to differentiate *in vivo* into endothelial cells necessary for tumor self-maintenance. Together, these data provide new insights for the development of novel drugs or therapies for treating glioblastomas that aim at inhibiting the endothelial transition of CD133[+] cancer stem cells.

The third speaker of the session was Catriona Jamieson (University of California, San Diego; Fig. 2). Her talk explored the molecular evolution of leukemia stem cells (LSCs) during progression of chronic myeloid leukemia (CML) from the chronic phase (CP) to blastic phase (BP). Several studies have shown that CML is characterized by the Philadelphia (Ph) chromosome, which has a BCR/ABL1 rearrangement with enhanced kinase activity.[18] Although it has been assumed that the BCR/ABL1 mutation occurs first in hematopoietic stem cells (HSCs) that are responsible for the development of the CML-CP phase,[19] little is known about the molecular mechanisms that trigger the disease to transition to the more severe CML-BP phase. Using RNA sequencing and nanoproteomics approaches, Jamieson identified splice variants and point mutations that commonly occur in all CML tumor samples processed. These candidate genes are grouped into three classes: (1) genes involved in LSC aberrant self-renewal (such as Jak2), (2) genes involved in LSC dormancy (Shh), and (3) genes that promote LSC survival (Bcl2). Using inhibitors against these genes, Jamieson presented preliminary *in vitro* results: (1) TG101348, a Jak2 inhibitor, when combined with desatinib, a BCR/ABL1 inhibitor, arrested stem cell self-renewal ability; (2) Shh inhibitors were able to reactivate the stem cell cycle transition from G0 to G1; and (3) sabutoclax, a pan-inhibitor of the Bcl2 family, induced apoptosis of LSCs, thus arresting the blast phase. Taken together, these preliminary results support the possible use of these target genes for development of novel therapeutics aimed at inhibiting CML-BP development.

Keynote address

The first invited keynote speaker was Irving L. Weissman (Stanford University; Fig. 3), whose work on hematopoiesis has been essential for defining the stages of development from hematopoietic stem cells (HSC) to their mature progeny. Many years of work on hematopoietic tissue, much of it led by Weissman's laboratory, has characterized phenotypically and genotypically nearly every step of differentiation of stem cells to mature progeny. Some of this basic science knowledge has been essential in understanding how normal hematopoietic cells become leukemic cells. For example, Weissman and colleagues have shown that despite preleukemic progression (such as EML1/ETO translocation) in HSCs, accumulation of further events in the progenitor compartment is responsible for the formation of leukemic stem cells (LSC).[20] Weissman's group identified the cell marker CD47, which is upregulated in mouse and human acute myeloid leukemia (AML) stem cells and functions as a "don't eat me signal." Weissman's group demonstrated that expression of CD47 on a human leukemia cell line improves tumor engraftment in immunodeficient mice due to evasion by CD47-expressing tumor cells of phagocytosis.[21] This suggests that cancer stem cells can defeat programmed cell removal by upregulating CD47 on their surface, which has led the way to a new anticancer treatment: blocking CD47 with an antibody so that phagocytosis can eliminate tumor cells, for example, AML stem cells.[22] Selective targeting of tumor cells by CD47 antibody is explained by the simultaneous presence on tumor cells, but not on most normal cells, of calreticulin, which acts as a prophagocytic signal.[23] Furthermore, the combination of CD47 antibody with rituximab, a monoclonal antibody that engages Fc receptors on NK cells and macrophages, has a synergistic action that results in total elimination of human non-Hodgkin lymphomas in mice.[24] Further studies on bladder, ovarian, and breast cancers suggest that different tumors use the same mechanism to escape immunosurveillance and can therefore be attacked by anti-CD47 treatment. This is a great example of translational research based on the evolution from basic science to the development of new drugs against human cancers.

Repairing our heart and muscles

The skeletal and cardiac muscle tissues constitute a large portion of the human body and are characterized by a set of different conditions that affect their

Figure 3. Irving L. Weissman (Stanford University and NYSCF Medical Advisory Board Member), Marc Tessier-Lavigne (The Rockefeller University and NYSCF Medical Advisory Board Member), Susan L. Solomon (CEO, New York Stem Cell Foundation), Mahendra S. Rao (Center for Regenerative Medicine, National Institutes of Health, and NYSCF Medical Advisory Board Member), and Craig B. Thompson (Memorial Sloan Kettering Cancer Center and NYSCF Medical Advisory Board Member) gather after a panel discussion entitled "The Future of Regenerative Medicine."

functionality. The understanding of the molecular mechanisms governing the development, regeneration, and repair of these tissues is fundamental for developing innovative therapeutic solutions for the treatment of a large variety of medical conditions associated with disability and death.

During her presentation Margaret Buckingham (Institut Pasteur) addressed the role of satellite stem cells in skeletal muscle regeneration, as well as the function of Pax genes and downstream targets in different stages of tissue development and regeneration. Satellite cells, quiescent cells found between the basement membrane and the sarcolemma of individual muscle fibers, undergo profound transcriptional changes upon activation. Microarray analysis of *in vivo* quiescent and activated cells reveals how satellite cells protect from oxidative damage, maintain quiescence, and are primed for activation when proper signals are provided.[25] This finding sheds light on the importance of extracellular matrix degradation for satellite cell migration and activation and demonstrates that upregulation of proteinases is crucial for optimal tissue regeneration *in vivo*. Buckingham's group has focused on the expression of Pax3 and its regulation of the myogenic determination factor Myf5 during development and regeneration. Genetic analysis and chromatin immunoprecipitation studies reveal that Pax3 specifically binds a conserved sequence upstream of Dmrt2 and regulates its expression. During tissue development, Dmrt2 regulates the early activation of the Myf5 gene (*Myf5*), which plays a central role in the formation of the first skeletal muscle in the somite.[26] In contrast, *Myf5* is already transcribed in satellite stem cells of adult muscle before the onset of myogenesis, although translation into a functional protein is repressed via binding of microRNA-31 and subcellular sequestration of the microRNA–RNA complex in ribonucleoprotein granules. Consistent with this expression of microRNA-31, antagonists of microRNA-31 in animal models of injury promoted tissue regeneration and increased fiber size. Upon activation, this post-transcriptional repression is released and cells undergo myogenic differentiation.[27] Buckingham highlighted the importance of silencing and sequestering tissue-specific regulatory gene transcripts to ensure rapid responses to tissue damage, and how similar control mechanisms may be involved in the regeneration of other somatic tissues.

Next, Deepak Srivastava (University of California, San Francisco) discussed his group's efforts to develop novel therapeutic strategies for human cardiac disorders based on known developmental pathways. By studying key molecular events during heart development, Srivastava's lab has identified a cascade of transcription factors and microRNAs that regulate early differentiation of cardiac progenitors and, later, their expansion into ventricular chambers.[28–36] Many of these pathways can be used to guide differentiation of pluripotent stem cells into cardiac, endothelial, and smooth muscle cells that may be useful for regenerative medicine. One example of translating these observations into regenerative approaches is the group's recent success in direct reprogramming of cultured mouse postnatal cardiac fibroblasts (CFs) into cardiomyocyte-like cells using a combination of three core developmental transcription factors, Gata4-Mef2C-Tbx5 (GMT).[37] These induced cardiomyocytes (iCMs) express cardiac-specific markers, have a global gene expression profile similar to neonatal cardiomyocytes, and exhibit spontaneous Ca^{2+} flux, electrical activity, and contraction. Compared to iPSC reprogramming, reprogramming of CFs to iCMs with GMT occurs more rapidly (starting at day 3) and with a higher efficiency, up to 20%. Furthermore, Srivastava's unpublished data confirm that this technique also works *in vivo*. Using a Cre-mediated lineage tracing technique, his group has demonstrated that resident CFs can be reprogrammed into iCMs through retrovirus-mediated GMT transduction of mouse heart. Interestingly, the *in vivo* reprogramming efficiency was greater than that observed *in vitro*; and iCMs exhibited properties, such as connexin 43 (Cx43) localization, ultrastructure under an electron microscope, action potential, and contractility, that closely resemble the native adult cardiomyocytes, suggesting a more complete reprogramming than *in vitro*. Most importantly, reprogrammed iCMs found in the scar zone formed after myocardium infarction resulted in improved cardiac function. Given that CFs normally compose over 50% of all the cells in the heart, these findings raise the possibility of reprogramming the vast pool of endogenous CFs into functional cardiomyocytes for regenerative purposes.

Michael Rudnicki (Ottawa Health Research Institute; Fig. 4) presented work on the identification of signaling pathways that regulate the function of

Figure 4. Michael Rudnicki (Ottawa Health Research Institute, Canada) presents his latest work with molecular regulation of muscle stem cell function.

satellite stem cells in adult skeletal muscle. Satellite stem cells support growth, homeostasis, and regeneration of skeletal muscle tissue through asymmetric (apical–basal) and symmetric (planar) divisions.[38] His laboratory has found that satellite cells uniformly express the transcription factor Pax7.[39] However, studies in mice have demonstrated that 10% of cells within the Pax7 population are negative for Myf5. Pax7[+]Myf5[−] cells were found to give rise to Pax7[+]Myf5[+] cells through asymmetric division within the satellite cell niche, indicating a positional control of cell fate. In agreement with this, transplantation studies revealed that the Pax7[+]Myf5[+] cells are a subpopulation of myogenic progenitors that preferentially differentiate, whereas Pax7[+]Myf5[−] cells contribute to satellite niche repopulation.[40] The ability of Pax7[+]Myf5[−] cells to expand the satellite pool suggests that these cells may have therapeutic potential in treating a large variety of degenerative disorders affecting the skeletal muscle system. Furthermore, Rudnicki described the mechanism through which Pax7 activity is regulated upon asymmetric cell division, resulting in Myf5 transcription. Tandem affinity purification and mass spectroscopy analysis led to the identification of a set of cofactors interacting with Pax7, including the Wdr5-Ash2L-MLL2 histone methyltransferase complex, which directs chromatin modification and allows the transcription of myogenic determination genes.[41] Similar studies have led to identification of additional Pax7 interacting proteins, such as protein arginine *N*-methyltransferase-4 (CARM1/PRMT4). CARM1 binds Pax7 and regulates its function, through methylation of N-terminal arginines, which is necessary for the recruitment of the Wdr5-Ash2L-MLL2 complex. Mutation of the methylation sites in the Pax7 sequence reduces the ability of Pax7 to upregulate Myf5 transcription, indicating a direct control of myogenic specification by CARM1 in the satellite compartment. Rudnicki also presented new findings demonstrating the common embryonic origin of brown fat and skeletal muscle. Isolated Pax7[+]Myf5[+] satellite cells were found to differentiate toward the adipogenic lineage when appropriately stimulated. The transcription factor PDRM16 was found to control the bidirectional fate switch between skeletal myoblasts and brown fat cells. Knockdown of PDRM16 in brown fat precursor results in increased expression of myogenic genes, whereas ectopic expression of PDRM16 in myoblasts induces brown fat differentiation. Furthermore, immunopurification studies followed by mass spectroscopy demonstrate that PDRM16 interacts with PPAR-γ,[42] recognized as a master transcriptional regulator of adipogenesis.[43] These data indicate a direct control of PDRM16 on brown fat differentiation.

Neurodegeneration and spinal cord injury

Steven Goldman (University of Rochester Medical Center), the first speaker of the session entitled "Neurodegeneration and Spinal Cord Injury," presented recent work on a new strategy to isolate human oligodendocyte progenitor cells (OPCs) from fetal human brain.[44] Following previously published data from his lab, in which OPCs were isolated by FACS based on the expression of the cell surface ganglioside marker A2B5, Goldman's group identified that OPCs can be further enriched by isolating cells expressing the receptor CD140a (PDGFRa), a subpopulation of the A2B5[+] cells.[45] CD140a[+] cells from human fetal brain, depleted of neuronal- and astrocytic-specific markers, are able to migrate and myelinate neuronal axons when transplanted into the hypomyelinated mouse brain. Myelination was faster and more efficient than that observed after transplantation of A2B5[+] cells that were not further enriched for CD140a expression. This finding holds true when human CD140a[+] cells are transplanted in chemically induced demyelinating rat brains and when OPCs are produced from human iPS cells, though with variable efficiency rates. Finally, Goldman presented interesting unpublished data from experiments with the shiverer mouse model following transplantation of CD140a$^+$ human OPCs into the mouse brains. The cells gave rise to myelinating oligodendrocytes and astrocytes, and the brains were found to be chimeric, with >70% of the glial cells being of human origin (8 months posttransplantation), suggesting that human donor cells outpace host-derived cells. Furthermore, the transplanted cells retained a human phenotype and function with high synaptic potentiation. Interestingly, Pavlovian-based behavioral tests show that the chimeric mice are better at learning by association and can be conditioned to fear faster than their wild-type counterparts, a sign that these mice might be "smarter." In addition to allowing the study of basic biological interactions of human glial cells, this strategy provides a unique mouse model in which to study diseases of the central nervous system, including human-specific infectious diseases.

In the second talk of the session, Paul Tesar (Case Western Reserve University) presented recently published work on the differentiation of mouse epiblast stem cells (EpiSCs) to OPCs.[46] Tesar and his group are attempting to produce pure populations of OPCs by directed differentiation, using the signals that normally play important roles during normal development of oligodendrocytes, and to identify the required developmental transitions to produce functional OPCs. In the first stage, pluripotent EpiSCs are specified to the neuroectodermal lineage and cells are organized into typical neural rosette structures by blocking the activin-nodal and BMP signaling pathways for four days. In the second stage, the neuroectodermal-like cells are patterned to region-specific neural precursor cells by addition of retinoic acid and SHH for one day. In the final stage, production of OPCs results from platelet-derived growth factor (PDGF) and FGF signaling after an additional five days. This ten-day differentiation protocol results in a population of highly enriched cells expressing OPC-specific markers that display the typical bipolar morphology of *in vivo* counterparts. Such mouse EpiSC-derived OPCs can be expanded for at least eight passages—results consistent across four independent EpiSC lines. After a further four days, the OPCs differentiate to highly specific mature oligodendrocytes (lacking any neuronal or astrocytic markers) that express basic myelin protein. The functionality of these OPCs was tested by injecting them into forebrain slices from the hypomyelinated shiverer mouse. In this context, the cells

produced oligodendrocytes that migrated and myelinated neuronal axons in the host mouse brain tissue. Tesar reported that similar experiments are under way to find analogous methodologies that will robustly produce OPCs from human iPS cells. Finally, Tesar presented data (submitted) involving direct conversion of fibroblasts to OPCs, with a 10–15% efficiency; the OPCs could expand, differentiate to mature oligodendrocytes, and myelinate sections of shiverer mouse brain.

Clive Svendsen (Cedars-Sinai Regenerative Medicine Institute) presented two approaches to treating Huntingon's disease (HD) and other neurodegenerative disorders. The first part of his talk summarized his lab's work over the previous 14 years using fetal brain tissue transplant approaches as a therapeutic intervention for HD and Parkinson's disease (PD) models.[47,48] This method includes isolation of human progenitors from the fetal cortex and then passage of the cells by a "chopping method" to insure cell–cell contact. After 20 passages, these progenitors can differentiate, at a high percentage, to astrocytes. Such astrocyte progenitors, injected into animal models, produce human astrocytes that take 120 days to mature. The therapeutic potential of these cells can be amplified using *ex vivo* gene therapy enabling the cells to produce the neuroprotectant glial cell–derived neurotrophic factor. This combination is effective in modulating the deleterious effects observed in the N171–82Q mouse model of HD1. Svendsen described attempts to move this approach to clinical application for a variety of neurologic disorders, including HD, ALS, macula degeneration, and stroke. The second part of Svenden's talk described the progress of the iPSC Stem Cell Consortium for Huntington's disease, which is a collaboration between investigators from multiple research institutions, including Svendsen's, to generate and characterize HD patient-specific iPSCs. This unpublished work, funded by an NIH Grand Opportunities Grant, provides an opportunity for analysis of the HD iPSCs by groups with different expertise, as well as independent validation of experimental results by different laboratories. iPSCs have been generated from patients with different levels of CAG repeats in the Huntington gene. The number of CAG repeats dictates whether a person will develop HD and positively correlates with age of onset.[49] Various labs within the consortium have demonstrated numerous disease-related phenotypes for neurons generated from multiple clones of higher CAG repeat iPSCs, including delays in neuronal maturation, adhesion properties, ATP metabolism, and cell death in response to stresses such as acute pulses of glutamate and BDNF withdrawal. Sensitivity to glutamate is also associated with decreased ability to reset calcium homeostasis, which may explain why HD neurons are more sensitive to excitotoxic insults.

NYSCF-Robertson Prize lecture

During the NYSCF–Robertson Prize lecture, Peter J. Coffey (University College London; Fig. 5) presented an overview of the work he has done with the London Project to Cure Blindness to find a treatment for age-related macular degeneration (AMD). AMD affects roughly 14 million older adults in the United States and in Europe. Currently, dry AMD, which affects 90% of the clinical population, has no known therapeutic treatment. Wet AMD, which affects 10% of the clinical population, can be treated with anti-VEGF therapy, but this therapy is expensive and time consuming, as patients must receive monthly injections of drugs into the eye.

In wet AMD, blood vessels, which perfuse blood into the retina, protrude through a weakened Bruch's membrane, the membrane dividing the choroid and the retinal pigment epithelial (RPE) cell layer.[50] The protrusion of these blood vessels into the RPE cell layer compromises its function and cuts off the nutrition and structural support it provides for the photoreceptors, eventually leading to detachment of the macula and vision loss.[51] In order to treat wet AMD, Coffey and colleagues have developed a macular translocation surgery in which the macula is reconstructed via its rotation. Macular translocation resulted in 25% of patients gaining three lines of acuity up to three years after surgery.[52] However, this surgery is very complex and time consuming, and the cost per patient is exceedingly high. Therefore, Coffey and colleagues attempted macular translocation with 360° retinotomy and autologous RPE–choroid patch graft, which has dramatically reduced costs as well as potential surgical complications, such as trauma to the photoreceptors themselves. However, RPE–choroid patch grafts were found to be inferior to macular translocation alone, partially because exogenous RPE grafts lacked sufficient metabolic support and the graft membrane did not allow sufficient attachment of the grated cells to the Bruch's membrane.[53] To address the problem of exogenous transplantation rejection, Coffey's group was first able to develop RPE cells generated from hESCs and successfully transplant these cells into the RCS rat model with inherited retinal degeneration. His group showed that photoreceptors could

Figure 5. Peter J. Coffey (University College London, the London Project to Cure Blindness, and recipient of the Inaugural NYSCF-Robertson Prize in Stem Cell Research) delivers a special keynote address on his use of human ESCs to cure age-related macular degeneration.

be rescued using both electrophysiological and behavioral tests, such as pattern discrimination and head tracking after implantation of hESC-derived RPE cells.[54,55] Indeed, grafts of RPE cells derived from iPS cells into RCS rats induce the short-term maintenance of photoreceptors and significantly increase visual acuity.[56] Coffey reported that experiments using choroid patch grafts from RPE-hESCs in pig retina result in the proliferation of functional photoreceptors with normal autofluorescence, indicating that the insertion of choroid patch grafts from RPE-hESCs into human eyes shows significant promise in the treatment of wet AMD. Phase I/II clinical trials based on this research are proposed for 2012.

Programming and reprogramming

Dieter Egli (New York Stem Cell Foundation; Fig. 6), the first speaker of the session entitled "Programming and Reprogramming," reported on two recent publications aimed at generating patient-specific stem cells through the use of somatic cell nuclear transfer (SCNT). Egli first highlighted growing concerns about iPS cells, stressing the need for continued work in SCNT.[57–59] However, many labs have failed to fulfill the potential of SCNT, thus posing the question of whether these failures are because of the preliminary nature of the studies, or because of an intrinsic issue with the human egg that prevents reprogramming.

Beginning with his experience at Harvard University, Egli highlighted the issues of recruiting oocyte donors when only "altruistic donors" can be enrolled.[60] After moving to the New York Stem Cell Foundation, Egli's study continued in collaboration with Columbia University Medical Center's Center for Women's Reproductive Care (CWRC), where, to date, 16 donors have participated. Initial experiments corroborated previous findings but also suggested that developmental failure was due to transcriptional dysfunction. Using microarray analysis, the expression profile of such failures closely resembled oocytes treated with transcriptional inhibitor aminitin, raising the following questions: (1) Is artificial activation of the oocyte causing this arrest? This was not the case, as parthenogenetic blastocysts could be made. (2) Is oocyte manipulation causing this arrest? This was answered *no* by demonstrating that if the genome was removed and replaced into the same oocyte, parthenogenetic blastocysts could still form. (3) Is the somatic cell genome interfering with early development, or is it the removal of the oocyte genome? To answer this, a somatic cell (GFP labeled) was fused to an oocyte with an intact oocyte genome. Following activation, development to the blastocyst stage demonstrated GFP activation and, hence, that the somatic cell genome had participated in blastocyst development. Stem cells derived from the blastocysts developed into all three germ layers and expressed markers typical of stem cells.

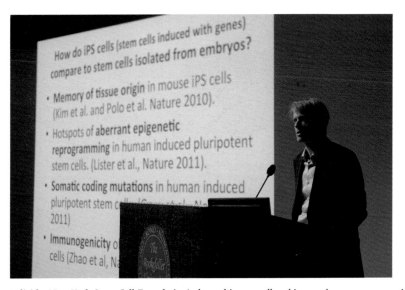

Figure 6. Dieter Egli (the New York Stem Cell Foundation) shares his groundbreaking work on reprogramming after nuclear transfer in the session entitled "Programming and Reprogramming."

However, these stem cells were karyotypically abnormal, with a triploid genome. To eliminate the issue of timing, the oocyte genome was removed at the first interphase, which led to developmental arrest. Sequencing analysis of both gDNA and cDNA was undertaken with an allelic ratio of ∼0.6 found in both, suggesting that the somatic genome was as active as the oocyte genome. Analysis of gene expression profiles of the stem cells showed that there was no preferential expression of either fibroblast or stem cell genes, suggesting that no "somatic memory" was present.[61] This work suggests that the oocyte can reprogram the somatic genome, and further work is now under way to identify what causes developmental arrest following removal of the oocyte genome.

Alex Meissner (Harvard University; Fig. 7) began his talk by referencing the same issues that Egli raised concerning iPS cells. While acknowledging that critical investigation is warranted, Meissner pointed out that there is also a wide range in variation among ES cells lines. These variations were shown by methylation mapping, gene expression profiling, and use of a quantitative differentiation assay. Furthermore, while some iPS cell lines do not cluster with ES lines in these analyses, many do, suggesting that they are as useful as many ES lines.[62] Meissner went on to discuss DNA methylation dynamics and how to further our understanding of where and when DNA methylation is used as a regulatory mechanism. He described unpublished data that highlights the critical role of DNA methyla-

tion during development. Zygotes from fertilized mouse oocytes were collected at multiple times during development. It was initially found that although there were few differences in global methylation patterns between E7.5 and adult tissue, analysis of the oocyte itself revealed a globally lower level of methylation than that in sperm. Following fertilization, a continuous shift in demethylation (of the paternal genome) is seen until E6.5 at implantation. Furthermore, the inner cell mass (ICM) of the blastocyst appears to differ greatly from that of the somatic genome in which classically low CpG dense areas are highly methylated. It was also found that for a number of genes, including *Dnmt1*, separate maternal and somatic methylation patterns exist. Until the eight-cell stage of development, the somatic *Dnmt1* CpG island is highly methylated before switching to a demethylated state, with the opposite being true for the maternal version. From unpublished, ongoing, collaborative work with Egli, the initial analysis of mouse nuclear transfer zygotes appears to show that nuclear transfer blastocysts show comparable global demethylation to that seen in zygotes. It appears, however, that similar to sperm, the oocyte is capable of removing some methylation from previously highly methylated areas, although a number of regions remain as methylated as in the somatic cell. Therefore, remodeling by the oocyte does take place; however, not all regions are remodeled equally.

Kevin Eggan (New York Stem Cell Foundation and Harvard University) ended the programming

Figure 7. Alex Meissner (Harvard University) discusses his progress in epigenetic reprogramming.

and reprogramming session by discussing his recently published work describing the direct conversion of mouse and human fibroblasts into induced motor neurons (iMNs) by a process referred to as *transdifferentiation.*[63] The purpose of this study was to generate amyotrophic lateral sclerosis (ALS) disease-specific motor neurons by transdifferentiation, which subsequently could be compared to similarly derived control motor neurons. Eggan, who pioneered the first ALS-specific iPS cells that could make disease-specific motor neurons,[64] suggested that iMNs, owing to their significantly faster generation time, could provide a quicker preview of disease-specific phenotypes than iPS approaches. Eggan's group first attempted to generate iMNs by virally overexpressing eight transcription factors alone, or in combination, known to be important for either the motor neuron differentiation process or in the mature motor neuron itself. This approach was unsuccessful in generating iMNs from fibroblasts obtained from HB9:GFP mice (which produce GFP-fluorescing motor neurons). However, when Eggan's group combined this strategy with the three transcription factors used by Marius Wernig's group to create more general neurons from fibroblasts (iNs),[65] they successfully generated iMNs. This could be reduced to a six or seven transcription factor cocktail, and could also be used to create iMNs from human fibroblasts. Similar to iNs,[66] the transdifferentiation process to iMNs does not appear to involve a neural progenitor intermediate. iMNs were shown to resemble bona fide motor neurons by a variety of criteria, including gene expression studies, electrophysiological properties, formation of cholinergic synapses with muscle cells generated from the C2C12 cell line, and implantation into the developing chick spinal cord with subsequent appropriate axonal migration to peripheral targets. Eggan finished his talk by showing preliminary unpublished data suggesting that human iMNs prepared from ALS patients with genetic forms of the disease demonstrate increased electrical excitability compared to control iMNs, which could predispose the former to increased sensitivity to excitotoxic insults.

Conclusion

NYSCF's sixth annual meeting convened leaders in stem cell research, both senior and innovative young scientists, providing attendees with novel scientific presentations that focused on a wide variety of diseases and techniques. The speakers, including two scientists from NYSCF's inaugural class of NYSCF-Robertson investigators, gave conference participants a first-hand look at unpublished and recently published pioneering research. The "Seventh Annual Translational Stem Cell Research Conference" will take place on October 10–11, 2012.

Acknowledgments

New York Stem Cell Foundation
Sixth Annual Translational Stem Cell Research Conference
Co-chairs
Lee Goldman, MD, MPH, Columbia University Medical Center
Antonio M. Gotto Jr., MD, DPhil, Weill Cornell Medical College
Douglas A. Melton, PhD, Harvard University
Allen M. Spiegel, MD, Albert Einstein College of Medicine
Marc Tessier-Lavigne, PhD, The Rockefeller University
Craig B. Thompson, MD, Memorial Sloan-Kettering Cancer Center
Scientific Co-chairs
Moses V. Chao, PhD, Skirball Institute of Biomolecular Medicine
Zach W. Hall, PhD, The New York Stem Cell Foundation
Ihor Lemischka, PhD, The Mount Sinai School of Medicine
Dan R. Littman, MD, PhD, Skirball Institute of Biomolecular Medicine
Principal Sponsor
Robertson Foundation
Co-sponsoring Institutions
Albert Einstein College of Medicine
Columbia University Medical Center
Helen and Martin Kimmel Center for Stem Cell Biology, New York University School of Medicine
Mount Sinai School of Medicine
Tri-Institutional Stem Cell Initiative:
- Memorial Sloan-Kettering Cancer Center
- The Rockefeller University
- Weill Cornell Medical College

Conflicts of interest

The authors declare no conflicts of interest.

References

1. Hall, Z.W. *et al.* 2010. Breaking ground on translational stem cell research. *Ann. N.Y. Acad. Sci.* **1189** (Suppl. 1): E1–E11.

2. Marshall, C. *et al.* 2011. The New York Stem Cell Foundation: Fifth Annual Translational Stem Cell Research Conference. *Ann. N.Y. Acad. Sci.* **1226**: 1–13.

3. Green, M.D. & H.W. Snoeck 2011. Novel approaches for immune reconstitution and adaptive immune modeling with human pluripotent stem cells. *BMC Med.* **9**: 51.

4. Green, M.D. *et al.* 2011. Generation of anterior foregut endoderm from human embryonic and induced pluripotent stem cells. *Nat. Biotechnol.* **29**: 267–272.

5. D'Amour, K.A. *et al.* 2006. Production of pancreatic hormone-expressing endocrine cells from human embryonic stem cells. *Nat. Biotechnol.* **24**: 1392–1401.

6. Chen, S. *et al.* 2009. A small molecule that directs differentiation of human ESCs into the pancreatic lineage. *Nat. Chem. Biol.* **5**: 258–265.

7. Thorel, F. *et al.* 2010. Conversion of adult pancreatic alpha-cells to beta-cells after extreme beta-cell loss. *Nature* **464**: 1149–1154.

8. Thorel, F. *et al.* 2011. Normal glucagon signaling and β-cell function after near-total α-cell ablation in adult mice. *Diabetes.* Epub.

9. Ding, B.S. *et al.* 2011. Endothelial-derived angiocrine signals induce and sustain regenerative lung alveolarization. *Cell* **147**: 539–553.

10. Butler, J.M. *et al.* 2010. Endothelial cells are essential for the self-renewal and repopulation of Notch-dependent hematopoietic stem cells. *Cell Stem Cell* **6**: 251–264.

11. Hooper, A.T. *et al.* 2009. Engraftment and reconstitution of hematopoiesis is dependent on VEGFR2-mediated regeneration of sinusoidal endothelial cells. *Cell Stem Cell* **4**: 263–274.

12. Kobayashi, H. *et al.* 2010. Angiocrine factors from Akt-activated endothelial cells balance self-renewal and differentiation of haematopoietic stem cells. *Nature Cell Biol.* **12**: 1046–1056.

13. Ding, B.S. *et al.* 2010. Inductive angiocrine signals from sinusoidal endothelium are required for liver regeneration. *Nature* **468**: 310–315.

14. Butler, J.M. *et al.* 2010. Instructive role of the vascular niche in promoting tumour growth and tissue repair by angiocrine factors. *Nat. Rev. Cancer* **10**: 138–146.

15. Singh, S.K. *et al.* 2004. Identification of human brain tumor initiating cells. *Nature* **432**: 396–401.

16. Wang, R. *et al.* 2010. Glioblastoma stem-like cells give rise to tumor endothelium. *Nature* **468**; 829–833.

17. Verhaak, R.G. *et al.* 2010. Integrated genomic analysis identifies clinically relevant subtypes of glioblastoma characterized by abnormalities in PDGFRA, IDH1, EGFR and NF1. *Cancer Cell* **17**: 98–110.

18. Cross, N.C. *et al.* 2008. BCR-ABL1-positive CML and BCR-ABL1-negative chronic myeloproliferative disorders: some commons and contrasting features. *Leukemia* **22**: 1975–1989.

19. Bruns, I. *et al.* 2009. The hematopoietic stem cell in chronic phase CML is characterized by a transcriptional profile resembling normal myeloid progenitor cells and reflecting loss of quiescence. *Leukemia* **23**: 892–899.

20. Miyamoto, T. *et al.* 2000. AML1/ETO-expressing non-leukemic stem cells in acute myelogenous leukemia with 8;21 chromosomal translocation. *Proc. Natl. Acad. Sci. USA* **97**(13): 7521–7526.

21. Jaiswal, S. *et al.* 2009. CD47 is upregulated on circulating hematopoietic stem cells and leukemia cells to avoid phagocytosis. *Cell* **138**(2): 271–285.

22. Majeti, R. *et al.* 2009. CD47 is an adverse prognostic factor and therapeutic antibody target on human acute myeloid leukemia stem cells. *Cell* **138**(2): 286–299.

23. Chao, M.P. *et al.* 2010. Calreticulin is the dominant pro-phagocytic signal on multiple human cancers and is counterbalanced by CD47. *Sci Transl Med* **2**(63): 63ra94.

24. Chao, M.P. *et al.* 2010. Anti-CD47 antibody synergizes with rituximab to promote phagocytosis and eradicate non-Hodgkin lymphoma. *Cell* **142**(5): 699–713.

25. Pallafacchina, G. *et al.* 2010. An adult tissue-specific stem cell in its niche: a gene profiling analysis of in vivo quiescent and activated muscle satellite cells. *Stem Cell Res* **4**(2): 77–91.

26. Sato, T. *et al.* 2010. A Pax3/Dmrt2/Myf5 regulatory cascade functions at the onset of myogenesis. *PLoS Genet* **6**(4): e1000897.

27. Buchan, J.R. *et al.* 2009. Eukaryotic stress granules: the ins and outs of translation. *Mol Cell* **36**(6): 932–941.

28. Garg, V. *et al.* 2003. GATA4 mutations cause human congenital heart defects and reveal an interaction with TBX5. *Nature* **424**: 443–447.

29. Bock-Marquette, I. *et al.* 2004. Thymosin beta4 activates integrin-linked kinase and promotes cardiac cell migration, survival and cardiac repair. *Nature* **432**: 466–472.

30. Garg, V. *et al.* 2005. Mutations in NOTCH1 cause aortic valve disease. *Nature* **437**: 270–274.

31. Zhao, Y. *et al.* 2005. Serum response factor regulates a muscle-specific microRNA that targets Hand2 during cardiogenesis. *Nature* **436**: 214–220.

32. Zhao, Y. *et al.* 2007. Dysregulation of cardiogenesis, cardiac conduction, and cell cycle in mice lacking miRNA-1-2. *Cell* **129**: 303–317.

33. Fish, J.E. *et al.* 2008. miR-126 regulates angiogenic signaling and vascular integrity. *Dev. Cell* **15**: 272–284.

34. Ivey, K. N. *et al.* 2008. MicroRNA regulation of cell lineages in mouse and human embryonic stem cells. *Cell Stem Cell* **2**: 219–229.

35. Cordes, K.R. *et al.* 2009. miR-145 and miR-143 regulate smooth muscle cell fate and plasticity. *Nature* **460**: 705–710.

36. Ieda, M. *et al.* 2009. Cardiac fibroblasts regulate myocardial proliferation through beta1 integrin signaling. *Dev. Cell* **16**: 233–244.

37. Ieda, M. *et al.* 2010. Direct reprogramming of fibroblasts into functional cardiomyocytes by defined factors. *Cell* **142**: 375–386.

38. Kuang, S. *et al.* 2007. Asymmetric self-renewal and commitment of satellite stem cells in muscle. *Cell* **129**(5): 999–1010.

39. Seale, P. *et al.* 2000. Pax7 is required for the specification of myogenic satellite cells. *Cell* **102**(6): 777–786.

40. Darabi, R. *et al.* 2011. Assessment of the myogenic stem cell compartment following transplantation of

Pax3/Pax7-induced embryonic stem cell-derived progenitors. *Stem Cells* **29**(5): 777–790.

41. McKinnell, I.W. *et al.* Pax7 activates myogenic genes by recruitment of a histone methyltransferase complex. *Nat. Cell Biol.* **10**(1): 77–84.

42. Rosen, E.D. *et al.* 1999. PPAR gamma is required for the differentiation of adipose tissue *in vivo* and *in vitro*. *Mol. Cell* **4**(4): 611–617.

43. Seale, P. *et al.* 2008. PRDM16 controls a brown fat/skeletal muscle switch. *Nature* **454**(7207): 961–967.

44. Sim, F.J. *et al.* 2011. CD140a identifies a population of highly myelinogenic, migration-competent and efficiently engrafting human oligodendrocyte progenitor cells. *Nat. Biotechnol.* **29**(10): 934–941.

45. Windrem, M.S., *et al.* 2008. Neonatal chimerization with human glial progenitor cells can both remyelinate and rescue the otherwise lethally hypomyelinated shiverer mouse. *Cell Stem Cell* **2**(6): 553–565.

46. Najm, F.J. *et al.* 2011. Rapid and robust generation of functional oligodendrocyte progenitor cells from epiblast stem cells. *Nat. Methods* **8**: 957–962.

47. Ebert, A.D. *et al.* 2010. Ex vivo delivery of GDNF maintains motor function and prevents neuronal loss in a transgenic mouse model of Huntington's disease. *Exp. Neurol.* **224**: 155–162.

48. Capowski, E.E. *et al.* 2007. Lentiviral vector-mediated genetic modification of human neural progenitor cells for ex vivo gene therapy. *J. Neurosci. Methods* **163**: 338–349.

49. Roos, R.A.C. 2010. Huntington's disease: a clinical review. *Orphanet. J. Rare Dis.* **5**: 40.

50. Simpson, E. 1986. Immune regulation. *Immunol. Today* **7**: 1.

51. Lu, B. *et al.* 2009. Long-term safety and function of RPE from human embryonic stem cells in preclinical models of macular degeneration. *Stem Cells* **27**: 2126–2135.

52. Chen, F.K. *et al.* 2010. Increased fundus autofluorescence associated with outer segment shortening in macular translocation model of neovascular age-related macular degeneration. *Invest. Ophth. Vis. Sci.* **51**(8): 4207–4212.

53. Chen, F.K. *et al.* 2009. A comparison of macular translocation with patch graft in neovascular age-related macular degeneration. *IOVS* **50**(4): 1848–1855.

54. Coffey, P.J. *et al.* 2002. Long-term preservation of cortically dependent visual function in RCS rats by transplantation. *Nat. Neurosci.* **5**(1): 53–56.

55. Gias, C. *et al.* 2007. Preservation of visual cortical function following retinal pigment epithelium transplantation in the RCS rat using optical imaging techniques. *Eur. J. Neurosci.* **25**(7): 1940–1948.

56. Carr, A.J. *et al.* 2009. Protective effects of human iPS-derived retinal pigment epithelium cell transplantation in the retinal dystrophic rat. *PLOS ONE* **4**(12).

57. Gore, A. *et al.* 2011. Somatic coding mutations in human induced pluripotent stem cells. *Nature* **471**(7336): 63–67.

58. Lister, R. *et al.* 2011. Hotspots of aberrant epigenomic reprogramming in human induced pluripotent stem cells. *Nature* **471**(7336): 68–73.

59. Ohi, Y. *et al.* 2011. Incomplete DNA methylation underlies a transcriptional memory of somatic cells in human iPS cells. *Nat. Cell Biol.* **13**(5): 541–549.

60. Egli *et al.* 2011. Impracticality of Egg Donor Recruitment in the Absence of Compensation. *Cell Stem Cell* **9**(4): 293–294.

61. Noggle, S. *et al.* 2011. Human oocytes reprogram somatic cells to a pluripotent state. *Nature* **478**(7367): 70–75.

62. Bock, C. *et al.* 2011. Reference maps of human ES and iPS cell variation enable high-throughput characterization of pluripotent cell lines. *Cell* **144**(3): 439–452.

63. Son, E.Y. *et al.* 2011. Conversion of mouse and human fibroblasts into functional spinal motor neurons. *Cell Stem Cell* **9**: 205–218.

64. Dimos, J.T. *et al.* 2008. Induced pluripotent stem cells generated from patients with ALS can be differentiated into motor neurons. *Science* **321**: 1218–1221.

65. Vierbuchen, T. *et al.* 2010. Direct conversion of fibroblasts to functional neurons by defined factors. *Nature* **463**: 1035–1041.

66. Qiang, L. *et al.* 2011. Directed conversion of Alzheimer's disease patient skin fibroblasts into functional neurons. *Cell* **146**: 359–371.

Ann. N.Y. Acad. Sci. ISSN 0077-8923

ANNALS OF THE NEW YORK ACADEMY OF SCIENCES
Issue: Annals *Meeting Reports*

Understanding chronic inflammatory and neuropathic pain

Jane P. Hughes,[1] Iain Chessell,[1] Richard Malamut,[2] Martin Perkins,[2] Miroslav Bačkonja,[3] Ralf Baron,[4] John T. Farrar,[5] Mark J. Field,[6] Robert W. Gereau,[7] Ian Gilron,[8] Stephen B. McMahon,[9] Frank Porreca,[10] Bob A. Rappaport,[11] Frank Rice,[12] Laura K. Richman,[13] Märta Segerdahl,[14] David A. Seminowicz,[15] Linda R. Watkins,[16] Stephen G. Waxman,[17] Katja Wiech,[18] and Clifford Woolf[19]

[1]MedImmune, Cambridge, United Kingdom. [2]AstraZeneca R&D, Montreal, Quebec, Canada. [3]LifeTree Research, Salt Lake City, Utah, and University of Wisconsin-Madison, Madison, Wisconsin. [4]University of Kiel, Keil, Germany. [5]University of Pennsylvania, Philadelphia, Pennsylvania. [6]Grünenthal GmbH, Aachen, Germany. [7]Washington University School of Medicine, St. Louis, Misssouri. [8]Department of Anesthesiology and Perioperative Medicine, Queen's University, Kingston, Ontario, Canada. [9]King's College London, London, United Kingdom. [10]The University of Arizona, Phoenix, Arizona. [11]U.S. Food and Drug Administration, Center for Drug Evaluation and Research, Division of Anesthesia, Analgesia, and Addiction Products, White Oak, Maryland. [12]Integrated Tissue Dynamics, LLC, Rensselaer, New York, and Albany Medical College, Albany, New York. [13]MedImmune, Gaithersburg, Maryland. [14]AstraZeneca, Södertälje, Sweden. [15]University of Maryland School of Dentistry, Baltimore, Maryland. [16]University of Colorado at Boulder, Boulder, Colorado. [17]Yale University School of Medicine, New Haven, and Veterans Affairs Connecticut Healthcare System, West Haven, Connecticut. [18]University of Oxford, Oxford, United Kingdom. [19]Children's Hospital Boston, Boston, Massachusetts

This meeting report highlights the main topics presented at the conference "Chronic Inflammatory and Neuropathic Pain," convened jointly by the New York Academy of Sciences, MedImmune, and Grünenthal GmbH, on June 2–3, 2011, with the goal of providing a conducive environment for lively, informed, and synergistic conversation among participants from academia, industry, clinical practice, and government to explore new frontiers in our understanding and treatment of chronic and neuropathic pain. The program included leading and emerging investigators studying the pathophysiological mechanisms underlying neuropathic and chronic pain, and experts in the clinical development of pain therapies. Discussion included novel issues, current challenges, and future directions of basic research in pain and preclinical and clinical development of new therapies for chronic pain.

Keynote Lecture

Clifford Woolf (Children's Hospital, Boston, Massachusetts) opened the conference with a keynote lecture on the importance of target selection. Woolf stressed that choice of a wrong target will guarantee failure, therefore successful development of novel analgesics is contingent on a detailed molecular understanding of the mechanisms of pain.[1] He reviewed new technologies which enable this process to shift from the standard hypothesis-based candidate approach used so far with very limited success, to an unbiased genome-wide strategy[2] a change to a discovery science strategy to which Woolf claims will reveal true novelty. This will include genome-wide association studies in patient cohorts, computational genetics in mice, proteomics, transcription profiling and genetic manipulation in model or-ganisms with iterative validation and replication.[3] While each individual strategy may have limitations, combining them increases confidence. To meet this challenge Woolf stressed the importance a cooperative pre-competitive effort, since he believes that no single investigator or company can accomplish this alone.

Session I: Novel Targets and preclinical discovery

Chaired by Iain Chessell (MedImmune, Cambridge, UK), the first session examined recent therapeutic breakthroughs based on small molecules and the emerging role of biologics as potential new therapies.

Frank Rice (Integrated Tissue Dynamics, Rensselaer, New York, and Albany Medical College, Albany, New York) opened the session by discussing

doi: 10.1111/j.1749-6632.2012.06561.x

Ann. N.Y. Acad. Sci. 1255 (2012) 30–44 © 2012 New York Academy of Sciences.

the role of plasticity in peripheral fibers and epidermal molecular organization. Rice summarized how normal tactile sensation, including acute pain, is perceived through predictable patterns of activity involving a mix of peripheral sensory neuron types whose cutaneous endings are differentially activated by particular physical properties of tactile stimuli. Various neuronal types supply Aβ fibers, Aδ fibers, or C fibers. Subtypes predictably terminate as endings that have a unique combination of morphology, disposition and molecular expression that impart unique albeit overlapping functional properties involving differing proportions and types of mechanical, thermal, and chemical stimuli. The normal variety, proportion and disposition of cutaneous innervation is genetically programmed through complex molecular interactions during development and sustained maintenance. Likewise, a programmed pattern of differential terminations develops in the central nervous system so that a given tactile encounter will produce a predictable pattern of neuronal activity whose correlations provide the basis of normal perception.

Rice demonstrated that the characteristics and distributions of sensory endings can be permanently profoundly altered in association with peripheral neuropathies as documented in numerous immunocytochemical studies particularly of post herpetic neuralgia (PHN), complex regional pain syndrome type 1 (CRPS1), and type 2 diabetic neuropathy (DN2) that have occurred naturally in humans and Rhesus monkeys. Chronic pain-related pathological changes also occur among the neurochemical properties of epidermal keratinocytes that normally participate in tactile sensory transduction and modulation. Importantly Rice also claims that different types of peripheral neuropathies can manifest different combinations of painful symptoms. For example, painful DN2 is commonly associated spontaneous burning pain especially occurring symmetrically in the hands and feet. By contrast, PHN is usually asymmetrical, limited to one dermatome, and typically manifests extremely sharp pain in response to normally non-noxious mechanical contact with the skin (mechanical allodynia) or excessive pain response to normally noxious stimuli (hyperalgesia). Such differences indicate that there are potentially different mechanisms of chronic pain that may warrant different disease-dependent therapeutic strategies.

Interestingly, under a variety of chronic pain conditions in humans, including PHN, CRPS1 and DN2, Rice stressed that a consistent finding has been a seemingly paradoxical loss of epidermal innervation that presumably would be the very nociceptors that would contribute to pain sensation. However, Rice showed that this paradox may be partly explained by electrophysiological detection of increased spontaneous activity and hypersensitivity of remaining cutaneous innervation. Although the neurons in the spinal cord and brainstem also have increased spontaneous activity and hypersensitivity, referred to as central sensitization, increasing evidence indicates that this is driven by aberrant peripheral input. Multi-molecular immunocytochemical analyses indicate that the depletion of epidermal innervation preferentially involves all types of innervation but is most severely a depletion of the C fiber innervation. Most remaining innervation has microfilament properties indicative of Aδ fibers, and their sensory endings can have aberrant branching and neurochemical properties that would be consistent with pathologically increased activity. The epidermal innervation is not only reduced but becomes unevenly distributed with clusters of endings separated by wide, uninnervated gaps. Importantly, Rice suggested that keratinocytes of the epidermis can also manifest an aberrantly increased expression of neural characteristics such as a beta-isoform of CGRP, which may be released, and voltage gated sodium channels that mediate keratinocyte ATP release. This may also contribute to increased activation of sensory endings that remain in the epidermis. For the future, Rice suggested that strategies to facilitate the ability to re-establish stable, predictable patterns of tactile neural activity, even if they are abnormal, may help alleviate chronic pain.

Stephen B. McMahon (King's College London, London, United Kingdom) next went on to discuss how most chronic pain states can be temporarily ameliorated by local anaesthesia of the affected tissues. He suggested that in these states, peripheral pain signaling pathways are being tonically activated. The mediators responsible are for the most part unknown but the limited efficacy of nonsteroidal anti-inflammatory drugs (NSAIDs) suggests the existence of factors other than prostanoids. One factor that has been identified and shown to act as a peripheral pain mediator is the neurotrophin nerve growth factor (NGF). McMahon reviewed a wealth of preclinical data and more recently clinical trial data has shown the efficacy of peripheral anti-NGF strategies in relieving several forms of chronic pain, most notably pain associated with osteoarthritis. McMahon also described a novel approach being pursued in his laboratory to identify other peripheral pain mediators.[4] His group undertook transcriptional profiling of human biopsy specimens from painful conditions to identify candidate mediators and then tested the role of these candidates and their mechanism of action in preclinical models. They used ultraviolet B (UVB) irradiation to induce persistent, abnormal sensitivity to pain in humans and rats. The expression of more than 90 different inflammatory mediators was measured in treated skin at the peak of UVB-induced hypersensitivity with custom-made polymerase chain reaction arrays. McMahon found a significant positive correlation in the overall expression profiles between the two species. The expression of several genes (interleukin-1β (IL-1β), IL-6, and cyclooxygenase-2 (COX-2)), previously shown to contribute to pain hypersensitivity, was significantly increased after UVB exposure, and there was dysregulation of several chemokines (CCL2, CCL3, CCL4, CCL7, CCL11, CXCL1, CXCL2, CXCL4, CXCL7, and CXCL8). Among the genes measured, CXCL5 was induced to the greatest extent by UVB treatment in human skin; when injected into the skin of rats, CXCL5 recapitulated the mechanical hypersensitivity caused by UVB irradiation. This hypersensitivity was associated with the infiltration of neutrophils and macrophages into the dermis, and neutralizing the effects of CXCL5 attenuated the abnormal pain-like behavior. McMahon's findings demonstrate that the chemokine CXCL5 is a mediator of some inflammatory pain states.

The emerging role of biologics as potential new therapies was then presented by Jane P. Hughes (MedImmune, Cambridge, United Kingdom). Hughes opened with the statement that discovery and development of novel pain therapies remains an imperative, but the ability to genuinely test the efficacy of novel therapies is often limited by effects at targets other than intended, particularly with novel small molecule approaches. As a result, few novel mechanisms are genuinely tested in the clinic with unequivocal evidence of pharmacological interaction with the target. In addition, Hughes discussed how the regulatory environment is increasingly challenging, with reasonable expectations for drugs with highly favorable safety profiles; despite the devastating effects on quality of life, co-morbidities and socioeconomic impact, chronic pain is considered to be non-life threatening, and thus the risk/benefit profile of novel therapies must meet very high expectations. Approaches that limit these off-target activities may provide a greater ability to genuinely test targets of choice clinically.

Hughes went on to suggest that biologic therapeutics, in particular monoclonal antibodies (mAb's) provide such an opportunity. She explained that the attraction of mAb therapeutics is several-fold. Antibodies provide excellent affinity and specificity of target recognition; rarely do mAbs interact with any target other than that selected. In addition, because of their relatively large size, in vivo stability, and their ability to be sequestered and recycled via interaction with the FcRn receptors on endothelial cells, they tend to have extended pharmacokinetic half lives (typically around 7–14 days), which reduces inter-dose frequency.[5] Together, Hughes suggested these properties increase the likelihood of achieving true target engagement, assuming the target is accessible, and thus testing new therapeutic targets without the associated difficulties of managing off-target interactions. However, it was made clear that there are drawbacks to mAb-based therapeutics; immunogenicity, while of low incidence with humanized mAbs,[6] can be an issue with long term administration.

When considering the use of mAbs for the treatment of pain, Hughes stressed that not only must the above benefits and limitations be taken into account, but consideration must also be given to the discoverability and developability of mAbs against various target classes. An optimal target for mAb intervention is a soluble mediator; cytokines, growth factors, and inflammatory mediators generally have distinct identities within their active regions, and outside of the CNS are localized in readily-accessible compartments such as blood or joints. There are numerous soluble factors implicated in the pathogenesis of pain which play a role in the peripheral system, including various cytokines (e.g., IL-6), prostanoid products and kinins, giving potentially an already wide choice of possible targets to investigate further.

Hughes also demonstrated that therapeutic mAbs have been successful in targeting cellular receptors where the ligand binding domain lies on the extracellular face; here a number of tyrosine kinase and T cell receptors have already been targeted by launched products for oncology and inflammatory indications. The extension here is utilization of mAbs to target G-protein coupled receptors, opening a wealth of target opportunities for the treatment of pain. Coupled with some advances in the targeting of ion channel targets with mAbs, Hughes claimed that the target landscape for analgesic biologics is indeed a broad one, even when considering only peripheral targets. Further development of blood-brain-barrier penetrating platforms will serve to open this landscape still further. Conventionally, mAbs also do not penetrate the blood-brain-barrier readily, with typical concentrations of mAbs within the CNS compartment after chronic dosing being around 0.1–0.5% of that of the systemic circulation.[7] However, Hughes showed that this appears to be a surmountable issue, demonstrating with examples such as antibody fusions to insulin, utilizing the insulin transporter to allow the construct to enter the brain.[8]

Stephen G. Waxman (Yale University School of Medicine, New Haven, Connecticut, and Veterans Affairs Connecticut Healthcare System, West Haven, Connecticut) discussed how sodium channels serve, within the mammalian nervous system, as obligate generators of the upstroke of nerve impulses. Sodium channel blocking drugs have been used, for many years, to treat epilepsy and related disorders with some success. And while the efficacy of existing sodium channel blockers for the treatment of chronic pain has been limited, a number of recent developments suggest, however, that this situation may change. Waxman summarized our current understanding of the nine different genes that encode nine distinct sodium channel isoforms (NaV1.1–NaV1.9), with different amino acid sequences, different physiological properties, and different distributions within the nervous system. Waxman suggested that for a number of years pain research has focused on the following question: are there sodium channels that are preferentially expressed within nociceptors—a pattern of distribution that might permit selective targeting—so that drugs with limited cardiac and/or CNS side effects might be developed?

Four sodium channel isoforms have emerged as attractive molecular targets in this regard.[9,10] NaV1.3 is up-regulated within DRG neurons following injury to their peripheral axons. According to Waxman, this sodium channel isoforms has a number of physiological attributes (rapid recovery from inactivation, robust response to small, slow depolarizations close to resting potential, and production of a persistent current) that increase cell excitability. But whether up-regulated expression of this channel within axotomized DRG neurons will allow selective targeting is currently under exploration.

NaV1.9, originally termed NaN, is specifically expressed within nociceptors, and does not appear to be present within any other types of nerve cells. As a result of its slow kinetics and broad overlap between activation and steady-state inactivation, this channel plays a strong role in modulating the excitability of nociceptors. Waxman explained that development of subtype-specific blockers of NaV1.9 have been limited by modest levels of expression of this channel isoform within heterologous expression systems, and the development of expression systems that will permit high-throughput screening remains a challenge.

Of other sodium channel isoforms, according to Waxman, NaV1.7 is preferentially expressed at high levels within dorsal root ganglion and sympathetic ganglion neurons and NaV1.8 is specifically produced within dorsal root ganglion neurons. These two channel isoforms work in tandem, with NaV1.7 acting as a threshold channel that responds to small, slow stimuli such as generator potentials by producing its own depolarization, bringing membrane potential closer to the threshold for activation of NaV1.8. NaV1.8 produces the majority of the inward membrane current underlying the action potential during repetitive firing of DRG neurons. Waxman mentioned that NaV1.7 has attracted special interest because of its key role in producing human pain syndromes. For example, gain-of-function mutations that enhance the activation of NaV1.7 produce inherited erythromelalgia, a striking disorder characterized by severe burning pain of the hands and feet.[11] On the other hand, gain-of-function mutations that impair inactivation of NaV1.7 produce another disorder, paroxysmal extreme pain disorder, characterized by perirectal, periorbital, and perimandibular pain. Finally, loss-of-function mutations of NaV1.7 produce channelopathy-associated insensitivity to pain, a disorder in which humans lack functional NaV1.7 channels and do not feel pain in response to stimuli or events that should be painful; individuals with this disorder display painless burns, painless fractures, etc. and undergo dental extractions and surgery without feeling pain.

It has also become clear during the past few years, said Waxman, that polymorphisms of sodium channel can modulate sensitivity to pain. The R1150W polymorphism of NaV1.7, for example, contributes to sensitivity to pain following nerve root compression, limb amputation, and in osteoarthritis. It is highly likely that other sodium channel polymorphisms will also be found to contribute to pain sensitivity. Waxman concluded by discussing studies on human subjects with NaV1.7 mutations that have demonstrated that some mutations can alter sensitivity of the mutant channels to sodium channel blocking agents such as mexiletine and carbamazepine.[12] Together with the recent observations on sodium channel gene polymorphisms, this suggests that the goal of genomically-based personalized pain pharmacotherapeutics may not be unrealistic.

Robert W. Gereau (Washington University School of Medicine, St. Louis, Missouri) next went on discuss how epigenetic modulation may be an important mechanism of analgesia. Several reports indicate that L-acetylcarnitine (LAC) can be considered as a therapeutic agent in neuropathic disorders including painful peripheral neuropathies. His studies aimed at defining the mechanism of LAC analgesia indicated that LAC acts in part by epigenetic regulation of mGlu2 metabotropic glutamate receptor expression. Consistent with this finding, Gereau showed that activation of mGlu2 is robustly analgesic in animal models, though the therapeutic utility of mGlu2 agonists for the treatment of pain is limited by the robust development of analgesic tolerance to mGlu2 agonists. Mechanistic studies suggest that LAC exerts this effect via regulation of p65/RelA acetylation. These findings led Gereau and colleagues to test whether inhibiting deacetylation using HDAC inhibitors might similarly have analgesic effects via upregulation of mGlu2. He found that indeed two separate HDAC inhibitors promote analgesia and upregulation of mGlu2, and that the analgesic effects of HDAC inhibitors are reversed by an mGlu2 antagonist. Importantly, they found no evidence for the development of analgesic tolerance on repeated dosing of HDAC inhibitors. Gereau concluded that "epigenetic" drugs

that increase mGlu2 receptor expression, including L-acetylcarnitine and inhibitors of histone deacetylases may have a unique analgesic profile with no tolerance to the therapeutic effect after repeated dosing.

The last speaker in the first session, Linda R. Watkins (University of Colorado at Boulder, Boulder, Colorado), described how her work over the past 18 years has challenged classical views of pain and opioid actions. Watkins summarized how glia (microglia and astrocytes) in the central nervous system are now recognized as key players in: pain amplification, including pathological pain such as neuropathic pain; compromising the ability of opioids, such as morphine, for suppressing pain; causing chronic morphine to lose effect, contributing to opioid tolerance; driving morphine dependence/withdrawal; driving morphine reward, linked to drug craving and drug abuse; and even driving negative side effects such as respiratory depression. It is well documented that glial activation arises under conditions of chronic pain from neuron-to-glia signaling. Intriguingly, Watkins suggests that the glial activation receptor that creates neuroinflammation under conditions of chronic pain is the same receptor that is activated by opioids. Atop this, Watkins stated that opioid effects on glia that create neuroinflammation are via the activation of a non-classical, non-stereoselective opioid receptor distinct from the receptor expressed by neurons that suppresses pain. According to Watkins, this implies that the effects of opioids on glia and neurons should be pharmacologically separable, leading to new drugs for the control of chronic pain increasing the clinical efficacy of pain therapeutics. As such, Watkins claims that drugs in development which target this glial activation receptor have shown efficacy as stand alone treatments for neuropathic pain, by blocking neuron-to-glia signaling, plus blocking unwanted side effects of opioids, as well as other drugs of abuse.

Session II: Transitioning from preclinical to clinical studies

The second session, moderated by Martin Perkins (AstraZeneca R&D, Montreal, Quebec, Canada) consisted of a series of presentations on the challenge of translating results between clinical and preclinical studies.

Frank Porreca (The University of Arizona, Tucson, Arizona) discussed the translational capacity of reflexive endpoints used in animal models of pain. The majority of animal models of experimental neuropathic pain rely on nerve injuries with consequential various behavioral manifestations believed to reflect the different mechanisms contributing to a pathological pain state (For review see Ref. 13). None of these models, however, capture affective dimensions of pain as they rely on, typically, reflexive evoked responses to external stimuli whereas pain patients complain, primarily, of ongoing pain independent of an external stimulus. There is, therefore, a need to evaluate spontaneous or *stimulus-independent* pain in animal models.[13–15]

Pain has a strong emotional component exemplified by its unpleasantness which may have a protective role.[16,19] Porreca hypothesized that chronic pain produces an aversive state providing behavioral motivation to seek relief that will be rewarding. This concept was explored in the conditioned place-pairing assay, in which pairing pain relief with a distinct context resulted in increased time spent in that context.[16] Importantly, such conditioned place preference (CPP) was only observed in rats with nerve injury, leading to the "unmasking" of spontaneous experimental neuropathic pain.[16–18]

Rats with experimental nerve injury were placed in boxes consisting of two chambers with different visual and textural characteristics. Following preconditioning, a vehicle treatment was paired with one chamber and spinal administration of clonidine or ω-conotoxin in the other chamber, resulting in place preference in these animals where they preferred this chamber where pain relief occurred, revealing the presence of spontaneous pain.[16] The place preference associated with pain relief in rats with SNL was prevented by lesioning the rostral anterior cingulate cortex (rACC)[18] consistent with studies indicating that the rACC mediates the affective component of evoked pain.[19–21] This approach was also used to demonstrate descending pain modulatory pathways are important in mediating nerve-injury induced spontaneous pain in rats with nerve injury following rostral ventromedial medulla (RVM) microinjection of lidocaine.[21]

Whether neuropathic pain results as a consequence of activity of injured or adjacent uninjured nerves has been controversial.[18] Peripheral nerve injury results in increased excitability and ectopic discharge in primary afferents, thought to contribute to pain (for review, see Ref. 22). However, the contribution of either injured or uninjured primary afferent bers, or both, in these processes remains unclear with various studies supporting injured and uninjured bers being important (for review, see [23]). Clinically it appears that injured bers are important in driving pain [24] but it has been difficult to demonstrate this clearly in animal models of axotomy.[25,26]

Porreca then described his studies to determine whether CPP could be demonstrated in animals with either partial or complete hind paw denervation to assess the role of injured bers in promoting spontaneous neuropathic pain. Selective place preference to either spinal clonidine or RVM lidocaine in animals was observed in rats with sciatic or sciatic/saphenous axotomy suggesting spontaneous pain arises from injured nerve fibers.

Further studies were done exploring the role of TRPV1 and/or NK-1 receptors. Systemic treatment with resiniferatoxin (RTX) that produces long-lasting desensitization of TRPV1 receptors blocked nerve injury-induced thermal, but not tactile, hypersensitivity as well as the place preference resulting from pain relief.[17] Ablation of NK-1 receptor–expressing cells in the spinal cord with a substance P–saporin construct blocked nerve injury thermal and tactile hypersensitivity as well as ongoing pain.[17] These data suggest that spontaneous neuropathic pain involves both TRPV1 and NK-1 mediated mechanisms.

Porreca concluded by stressing the need for increased exploration of mechanisms associated with ongoing pain to improve translation from animal studies to human therapeutics leading to improved treatment of pain in humans.

Mark J. Field (Grünenthal GmbH, Aachen, Germany) then discussed the Pharma industry's productivity crisis with ever-increasing costs for research and development and decreasing approvals for new medicines. This is evident in the pain field with few new medicines successfully moving through development to the market to help patients. Over the past 10–15 years compounds with novel mechanisms of actions such as pregabalin, duloxetine and most recently tapentadol have made it to the market but still a large proportion of patients remain refractory to the available treatments. Field stressed that the translation of positive preclinical data into clinical efficacy is a key area of focus as many promising compounds / mechanisms fail to reach a positive proof of concept study. A major issue is the lack of understanding of the complex clinical chronic pain condition and the simple numeric rating scales used to assess clinical pain that give limited feedback to preclinical scientists. The concept of *translational research* hopes to build bridges between preclinical scientists and clinicians and identify biomarkers to assist in the development of novel pain medicines. Biomarkers can range from complex imaging techniques to simple blood borne markers that allow assessment of target engagement, pharmacology and even efficacy in early clinical trials. Closer co-operation between preclinical and clinical scientists through open innovation and pre-competitive consortia will be essential to ensure greater success in identifying novel medicines to treat the patients.

David A. Seminowicz (University of Maryland School of Dentistry, Baltimore, Maryland) discussed rodent brain imaging and behavior, stressing the importance of neuroimaging studies in humans, because perceptions and behaviors are results of brain processes. We strive to understand brain mechanism associated with acute and chronic pain. There is great potential for brain imaging leading to the identification of brain areas and functional networks that can be targeted to treat chronic pain. Neuroimaging in rodents can be used to supplement human imaging studies by addressing specific questions requiring animal models. For example, in rodent imaging studies we can examine the effects over time of an injury producing long-lasting pain behaviors, and we can perform histological studies examining the cellular correlates of brain imaging results and cellular correlates. Other advantages of neuroimaging in rodents include having tight control over genetic and environmental variables as well as performing invasive interventions such as drug injections and brain lesions. Two major disadvantages, however, for animal neuroimaging are that animals are usually anesthetized during the scan, which alters brain function, and we cannot accurately assess a brain-behavior relationship the way we can in humans by administering pain and asking subjects to rate its intensity.

There is extensive neuroimaging evidence that people with chronic have altered brain function and structure, the most common finding being a reduced gray matter volume or density or cortical thickness in widespread brain areas. Yet, two questions remained: (1) are these brain changes the cause or the consequence of chronic pain; and (2) are these brain changes reversible? Seminowicz and colleagues addressed the first question in a rodent model of neuropathic pain using a longitudinal design to follow the onset and progression of brain changes.[27] They examined brain structural changes and behaviors in rats with the spared nerve injury (SNI) model of peripheral neuropathic pain and found that in areas related to the sensory aspect of pain, including the anterior cingulate cortex and primary somatosensory cortex, decreased volume correlated with mechanical hypersensitivity. Furthermore, the volume of the prefrontal cortex decreased several months after the injury and corresponded in time with signs of anxiety-like behavior. They thus dissociated changes in sensory-related areas (S1, ACC), and affect-related areas (PFC) (Fig. 1). An important finding was that the PFC changes occurred only several months after the onset of nerve injury. Most experimental neuropathic pain studies only study animals for days or weeks after injury and thus may miss important aspects of behavioral changes associated with long-term pain. Finally, terminal histological analysis on the brains—about six months after SNI—revealed astrocyte proliferation in the PFC region with volumetric decreases but no obvious changes in neuronal or glial morphology or density.

Seminowicz highlighted some of the recent advances and remaining challenges in the field of neuroimaging in rodents including imaging awake animals, magnetic resonance spectroscopy, real-time fMRI, manganese enhanced MRI, and spinal cord imaging. A major challenge will be linking behavior and brain function in rodents, and translating these correlations to humans. He also described ongoing work with Radi Masri at the University of Maryland Dental School, including studies on electrical stimulation, resting state fMRI, and awake imaging.

Finally, to address the second question—i.e. whether the brain changes seen in chronic pain are reversible—Seminowicz discussed recent human neuroimaging studies suggesting that anatomical changes are reversible,[28–30] as well as his own work suggesting both structural changes and abnormal cognitive-related functional activity are reversible with effective treatment.[31] Together, the imaging

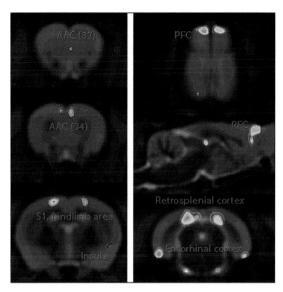

Figure 1. Sensory-related areas: volume decreases in these areas correlated with increased mechanical sensitivity (left panel). Affect-related areas: volume decreases in these regions corresponded in time with the onset of anxiety-like behavior (right panel).

data from animal and human studies suggest that pain causes anatomical brain changes and that treatment leads to recovery of normal brain anatomy.

In the first part of her presentation, Katja Wiech (University of Oxford, Oxford, United Kingdom) outlined the motivation to search for an objective readout for pain. Despite its enormous clinical relevance and socio-economic impact, pain and its representation of pain in the brain are only partly understood. Given the pivotal role of the brain for the perception of pain, there is an increasing demand for more detailed insights that could aid the development of new treatments. There is also a need for an objective indicator of pain to guide treatment in clinical settings. Patients' subjective reports are still the gold standard for the assessment of relevant pain features such as intensity or unpleasantness. A more objective measure is desirable if individuals are unable to provide ratings as with unconscious patients or preverbal infants. Finally, there is a growing interest in the use of neuroimaging in compensation claims related to persistent pain following work-related injuries as an objective marker for the presence or absence of pain could help determine the legitimacy of such claims.

Wiech then discussed the conceptual and methodological obstacles hampering the identification of such a biomarker for pain in the brain and recent advances in overcoming these challenges. It is now widely acknowledged that the perception of pain is not a simple reflection of incoming sensory information but a highly subjective experience that is determined by sensory as well as affective and cognitive factors. Hence, measures of pain that could serve as a biomarker has to consider the multidimensionality of pain. While the conventionally adopted univariate analysis approach has proven powerful for inference on highly localized structure-function mappings in the brain, it is insensitive to spatially distributed patterns of neural activity that are characteristic for highly complex phenomena such as pain. Multivariate pattern analysis (MVPA) is an emerging technique in functional brain imaging allowing for the integration of information from an extended network of brain regions.[32] Multivariate decoding models, such as those underlying classification algorithms, can be used to infer, directly, a perceptual state from the activity pattern across voxels—an application that has only recently been implemented in the context of pain.[33,34] In these first studies a classification algorithm was trained to differentiate between several intensity levels of pain induced by different stimulation intensities. The results show that, in principle, decoding of pain from brain images is feasible with prediction accuracies significantly above chance level.

Other commonly used indirect measures of brain activity such as blood oxygenation level dependent (BOLD) contrasts are partly ill-suited for the investigation of prolonged pain states due to methodological constraints such as the reduced sensitivity to events exceed a duration of approximately one minute. The development of fMRI-based arterial spin labeling (ASL) that measures changes in cerebral blood flow directly is less prone to time-sensitive effects and has opened up new possibilities to investigate chronic pain for which objective measures are urgently needed with promising results from the first studies in prolonged pain states.[35,36]

Results of functional neuroimaging studies on pain are commonly based on group data. Conclusions on individual cases as, for instance, required in the legal context, are often compromised by a relatively low signal-to-noise ratio (SNR). Ultra high field MR systems that are increasingly available could provide a better SNR due to increased sensitivity to the BOLD effect allowing for smaller sample sizes or even the identification of an objective measure for pain in the individual.

Märta Segerdahl (AstraZeneca, Södertälje, Sweden) discussed how abundant human pain models have been developed to understand pain physiology. Models range from assessments in normal skin to models inducing increased sensitivity to pain upon provocation, i.e. evoked pain. Lately also models of ongoing pain have been characterized. Pain has been induced by ischemia, heat, UV irradiation, chemical agents, electricity, etc. Most models have been pharmacologically validated with well-known analgesics, such as opioids, ketamine and non-steroidal anti-inflammatory drugs (NSAIDs), and have been used in drug development, mainly in single dose settings. Advantages are the possibility to conduct small studies with low variability under controlled conditions. However, their role in predicting later stage efficacy has been disappointing. Is there then still a role for such models in drug development? When developing new analgesics for new and unprecedented targets, there may be a place for these models. Models involving well understood mechanisms of action, such as capsaicin injection have been used for candidate drug selection for TRPV1 compounds, as a first step in demonstrating human targets engagement. Lately, models of inflammation, such as UV irradiation, have been characterized, linking models to known and new targets. There is a great advantage in being able to build confidence in the target mechanism by local testing avoiding systemic exposure, before traditional first time in man safety and tolerability studies. In this respect, these models well deserve their place in the drug development process of new analgesics.

Laura K. Richman (MedImmune, Gaithersburg, Maryland) began her talk by stating that chronic pain remains a significant problem that has few effective therapies. Choosing the right patients to receive a targeted therapeutic is critical from both the efficacy and safety perspective. There is a great need to reduce the cost of developing new pharmaceuticals, now approaching 1 billion per drug. The bulk of this cost is attributed to failed drugs. The FDA estimates that a 10% improvement in predicting clinical trial failures could reduce the average cost of drug development by nearly $100 million. An effort to generate more discriminatory biomarkers of efficacy and toxicity should eliminate suboptimal compounds earlier in development. Traditional preclinical animal studies often do not predict human-specific metabolic and toxic effects. Currently, the most commonly used toxicity biomarkers generate safety signals only when substantial organ damage has occurred. Ongoing efforts at finding efficacy and safety biomarkers that can predict or anticipate either desirable or undesirable effects in chronic pain and other indications were discussed.

Session III: Post-candidate clinical development

The third session was chaired by Richard Malamut (AstraZeneca R&D, Wilmington, Delaware) and focused on the challenges of clinical development of analgesic treatments.

Miroslav Bačkonja (LifeTree Research, Salt Lake City, Utah, and University of Wisconsin-Madison, Madison, Wisconsin) began the session with a presentation summarizing the current standard in the treatment of pain. Backonja noted that chronic pain is a complex disorder, manifesting with persistent pain and many other somatosensory symptoms, disturbance of sleep and mood. It results in negative impact on work, function and quality of life. He went on to state that both multi-modal and multidisciplinary approaches have been accepted and promoted by professional pain societies and organizations as the standard for managing chronic pain. Elements of multimodal approach to pain management include pharmacotherapy, interventional treatments, physical medicine and rehabilitation, psychological counseling and the development of appropriate coping skills. Drugs proven in randomized clinical trials to be effective in providing partial pain relief are the basis of pharmacotherapy of pain. These drugs are most commonly used in combination thereby taking advantage of diverse mechanisms of action. Interventional approaches include injections at anatomic sites presumed to be sources of pain and usually in combination with local anesthetics and steroids. Backonja then moved beyond pharmacologic approaches for treatment of pain, stating that neurostimulation therapy with implanted devices are frequently used as a last resource for pain relief. Physical and rehabilitation medicine treatments include a wide range of physical modalities but most critical is the patients' engagement into an active exercise program. Psychological therapies are important for the development of strategies to diminish both psychological co-morbidities and catastrophizing as well as for the development of coping skills that allow patients to deal with chronic pain more efficiently. In conclusion, Backonja firmly stated that an optimal multimodal approach is best achieved when trained clinicians from all disciplines, equipped to administer their respective therapies, work together in coordinated fashion.

Ralf Baron (University of Kiel, Kiel, Germany) then spoke about novel ways of segmenting patients with pain and the potential impact upon both clinical studies and clinical practice. In the past the classification of chronic pain syndromes had been based on disease entities, anatomical localization or histological observations. Over the past decade there has been a dramatic increased understanding of the pathophysiological mechanisms leading to the generation of chronic pain. Exciting advances in basic science have occurred in parallel with a growing awareness by clinical investigators that chronic pain is not a monolithic entity, but rather presents as a composite of different pain qualities and other sensory symptoms. Baron believed that the traditional classification of chronic pain may be supplemented with a new classification in which pain is analyzed on the basis of underlying neurobiological mechanisms rather than on the basis of the etiology.[37,38] This mechanism- or symptom-based classification includes pain symptoms such as burning or shooting sensations as well as negative and positive sensory signs. He reviewed his innovative research in which the characteristic profile of sensory symptoms (a combination of negative and positive signs) can be elucidated in each patient by utilizing a battery of several standardized quantitative sensory tests (QST). Verbal descriptors from validated questionnaires can then depict the quality and intensity of the individual pain. Baron is a member of the German Research Network on Neuropathic Pain which has established a large data-base of > 2000 patients with diverse neuropathic pain states that includes epidemiological and clinical data as well as standardized quantitative sensory testing (QST).[39] He informed the audience that epidemiological and clinical data on the symptomatology of 4200 patients with painful diabetic neuropathy, postherpetic neuralgia and radicular pain from a cross sectional survey (painDETECT) is also available (Fig. 2).[40]

Baron went on to say that within disease entities, different subgroups of patients can be distinguished on the basis of the individual sensory profile (sensory phenotype). These subgroups are present across etiologies but occur in different frequencies. By comparing the sensory patterns of human surrogate pain models (e.g., the capsaicin model or the menthol model) with patient subgroups, more information can be learned about the underlying pain mechanisms that operate in chronic pain.

The final question posed by Baron was whether the different phenotypes (which are presumably related to different mechanisms) are associated with different treatment outcomes. To date several small QST trials have been performed to identify predictive factors of the response to medical treatments. A retrospective analysis of the treatment response in phenotypic subgroups of patients in large clinical trials could demonstrate a differential effect of the study medication between these subgroups.

In summary, Baron stated that this approach of classifying and sub-grouping patients with chronic pain on the basis of symptoms or sign provides the opportunity to stratify patients. The study population can be enriched prospectively in proof of concept studies on the basis of *a priori* defined entry criteria. Enrichment with patients who may be predicted to respond to a specific treatment should increase the likelihood for positive trial outcomes. The implications for clinical practice are an increased capability to design an individualized therapy, i.e. to

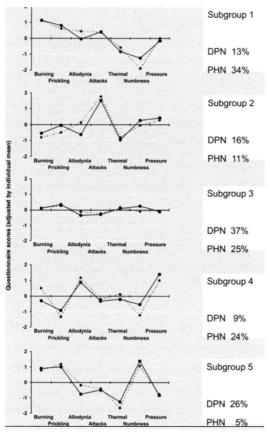

Figure 2. Subgrouping of patients according to sensory profiles using patient reported outcomes (PainDETECT questionnaire). To identify relevant subgroups of patients who are characterized by a characteristic symptom constellation a hierarchical cluster analysis was performed in a cohort of 2100 patients with painful diabetic neuropathy and postherpetic neuralgia. The clusters are represented by the patterns of questionnaire scores, thus showing the typical pathological structure of the respecting group. By using this approach five clusters (subgroups) with distinct symptom profiles could be detected. Sensory profiles show remarkable differences in the expression of the symptoms. % = frequency of occurrence, DPN = diabetic painful neuropathy, PHN = postherpetic neuralgia. Adjusted individual mean: in order to eliminate inter-individual differences of the general perception of sensory stimuli (differences individual pain perception thresholds) a score was calculated in which the given 0–5 score of each question was subtracted by the mean of all values marked in the seven questions. In this individual score values above 0 indicate a sensation which is more intense than the individual mean pain perception, values below 0 indicate a sensation which is less intense than the individual mean pain perception. From Baron *et al.*[4]

identify the patients who would be predicted to respond best to a specific treatment option.

Ian Gilron (Queens University, Kingston, Ontario, Canada) spoke of the challenges of designing clinical studies in neuropathic pain which will provide both valid and reliable results in order to identify beneficial interventions and will provide rigorous evidence which help patients and their caregivers to balance treatment benefits with the costs and risks of those interventions.[41] Optimal design of neuropathic pain clinical trials requires careful consideration of the specific research goals intended which is often dictated by the phase of drug development. Gilron then went on to discuss four components that contribute to a clinical study in patients with pain: study treatments, study populations, outcome measures, and trial designs.

Study treatments. With respect to pharmacological investigations, Gilron stated that design of clinical trials must carefully consider the optimal route of administration, duration of treatment, dosage formulation, and dose amount/frequency as determined by prior pharmacokinetic evaluations (for example, see Ref. 42). The use of placebos requires careful ethical consideration and, in situations where placebos are not appropriate or valid, alternatives to a placebo-controlled design are sometimes considered such as dose-ranging studies.[43,44] Active placebos (e.g., non-analgesic drugs mimicking study drug side effects) have been used in order to improve subject blinding to treatment.[45]

Study populations. Gilron explained that defining the study population is extremely important to the generalizability of trial results as well as to feasibility and conduct of the trial. He defined important considerations which included methods to recruit patients, diagnostic categories (e.g., the broad *peripheral neuropathic pain* vs. a specific entity such as postherpetic neuralgia), use of pain mechanism classifications, such as presence of allodynia,[46,47] disease/pain duration, pain severity/variability, concomitant analgesics and response to prior treatment. He stated that future research is needed to determine if and how these characteristics should dictate eligibility into various trials.[48]

Outcome measures. Gilron communicated that, while measures of pain intensity are the most commonly used primary outcome in neuropathic pain trials, there has been a growing appreciation that chronic pain profoundly impairs various aspects of quality of life. This has led to the development of IMMPACT (the Initiative on Methods, Measurement, and Pain Assessment in Clinical Trials) consensus recommendations that various multiple core domains,[49] and specific measures for these domains,[50] be considered in clinical pain treatment trials including: (1) pain, (2) physical functioning, (3) emotional functioning, (4) participant global ratings of improvement and satisfaction with treatment, (5) symptoms and adverse events, and (6) participant disposition.

Trial designs. Gilron noted that methods for allocating study subjects to various treatment conditions within a trial are critical to the conduct and interpretation of trial results.[41] In a parallel groups design, enrolled trial participants are randomized to one of two or more treatment groups (i.e., study treatment, placebo, active comparator). Study drug may be evaluated as monotherapy or as adjunct to existing analgesic therapy. In a crossover design, subjects are randomized to one of two or more sequences of treatment groups (i.e., study treatment, placebo, active comparator).[51] Crossover designs are considerably more powerful and efficient because each subject serves as his/her control that thus eliminates many sources of variability. However, Gilron cautioned that a crossover design requires stability of the underlying condition throughout all treatment periods, need to account for the potential of *carryover effect* from one treatment period to the next and may require a washout period.

Gilron concluded by describing potential future directions in neuropathic pain study design. He summarized that investigators continue to grapple with several ongoing challenges and priorities including: (1) internal validity (freedom from bias), (2) external validity (generalizability), (3) assay sensitivity (reduced variability/better precision), and (4) feasibility of subject recruitment/retention. Ongoing efforts to address these challenges include: (1) proposals for innovation in pain measurement methods (e.g., identifying the ideal primary outcome, electronic data capture, optimal timing & frequency,[52] (2) systematic review of clinical trial characteristics with a view to evidence-based trial design,[53] (3) trial design modifications for the improvement of assay sensitivity,[41,48] advancement of academia-industry-regulatory collaborations to strengthen analgesic research.[54] But as a final message suggesting hope for the future of developing better medications for patients, Gilron firmly stated that with the growing interest and effort into the improvement of various aspects of clinical trial design, it is anticipated that promising new treatments will be more optimally evaluated with respect to costs, risks and the healthcare benefits that they may provide to individuals suffering from neuropathic pain.

John T. Farrar (University of Pennsylvania, Philadelphia, Pennsylvania) provided a nice segue from the previous presentation in his discussion of outcome measurements in analgesia studies. He noted that the subjective nature of pain and pain measurements have often been sighted as a reason why clinical trials of potential therapies are difficult to conduct and interpret. However, beginning in the middle of the last century with simple parallel, two group, short term, randomized trials of analgesics (by Ray Houde, Henry Beecher, and others) consistent and valid results have been obtained for a variety of analgesics including non-steroidal anti-inflammatories, opioids, and adjuvant therapy. The underlying principals of random allocation, blinding, and a priori hypotheses based on primary pain outcomes remain the underpinnings necessary to produce valid clinical trial results. Standardization of patient reported pain outcomes (including pain intensity, pain interference, and global perception of change) have been validated in hundreds of studies. Farrar highlighted that an expanded conceptual framework for the pain process has provided a greater level of understanding of physiologic components underlying the pain experience and how they influence self reports of pain. There has been an improved understanding of the appropriate use of measures and analysis techniques and how to provide clinically useful interpretations of pain studies. Farrar went on to state that an improved understanding of the physiologic processes that underlie the pain response have also resulted in the development of a number of new pain therapeutics which underscores the need to create standard approaches to pain studies to allow for comparisons across disease states and types of therapies. He closed by confirming for the audience that improved efficiency of these studies has, in fact, become an important focus of current research.

Bob A. Rappaport (U.S. Food and Drug Administration (FDA), Center for Drug Evaluation and Research (CDER), Division of Anesthesia, Analgesia, and Addiction Products (DAAAP), White

Oak, Maryland) was the final speaker of the session. He discussed regulatory considerations in the development of analgesic medications. Rappaport stated that the FDA has long been concerned about the paucity of safe and effective analgesic drug products. To rectify this problem, the FDA has worked closely with the academic community and pharmaceutical industry to better understand the impediments that have led to a limited supply of novel analgesic drugs in the development pipeline. Representatives from the FDA have worked with the IMMPACT consensus organization for nearly a decade to provide recommendations for improving the design, conduct, and analysis of clinical studies of analgesic treatments. The clinical pain team in DAAAP has interacted with numerous stakeholders over many years to develop a guidance document for industry that will clarify the regulatory requirements for analgesic drug product development. Recent efforts to assure the continued availability of certain potent opioid drug products have included the development of risk, evaluation, and mitigation strategies designed to assure the safe use, storage, and prescribing of these products. Rappaport informed the audience that recently the FDA has taken the proactive step of creating a public–private partnership (ACTION) under the Critical Path Initiative, which is intended to advance the field of analgesic drug development at a more rapid pace by bringing together the best minds to address the problems associated with clinical studies of analgesics, and to find the funding necessary to perform the research to achieve that goal.

Summary

The conference "Chronic Inflammatory and Neuropathic Pain" provided an excellent environment for lively, informed, and synergistic conversation among participants. Although a great deal of superb science is taking place, if the overwhelming goal of research is to improve patient health, the conference highlighted the need for a cooperative, precompetitive effort from academia, industry, clinical practice, and government to explore new frontiers in our understanding and treatment of chronic pain.

Acknowledgments

The conference "Chronic Inflammatory and Neuropathic Pain," jointly presented by the New York Academy of Sciences, MedImmune, and Grünenthal GmbH, was supported in part by NYAS Bronze Sponsor Depomed, Inc., and by NYAS Academy Friend Sponsors Bristol-Myers Squibb Research and Development, and Regeneron Pharmaceuticals, Inc.

Conflicts of interest

Jane Hughes, Iain Chessell, and Laura K. Richman are employed by MedImmune. Richard Malamut, Martin Perkins, and Märta Segerdahl are employed by AstraZeneca.

References

1. Costigan, M., J. Scholz & C.J. Woolf. 2009. Neuropathic Pain: A Maladaptive Response of the Nervous System to Damage. *Annual Review of Neuroscience* **32:** 1–32.
2. Neely, G.G., A. Hess, M. Costigan, A.C. Keene, *et al.* 2010. A genome-wide Drosophila screen for heat nociception identifies α2δ3 as an evolutionarily conserved pain gene. *Cell* **143:** 628–638
3. Woolf, C.J. 2010. Overcoming obstacles to developing new analgesics. *Nat Med* **16:** 1241–1247.
4. Dawes, J.M., M. Calvo, J.R. Perkins, *et al.* 2011. CXCL5 mediates UVB irradiation-induced pain. *Sci Transl Med.* **3**(90): 90ra60.
5. Daugherty, A.L. & R.J. Mrsny. 2006. Formulation and delivery issues for monoclonal antibody therapeutics. *Adv. Drug Del. Rev.* **58:** 686–706.
6. Filpula, D. 2007. Antibody engineering and modification technologies. *Biomol. Eng.* **24:** 201–215.
7. Bacher, M., C. Depboylu, Y. Du, C. Noelker, *et al.* 2009. Peripheral and central biodistribution of (111)In-labeled anti-beta-amyloid autoantibodies in a transgenic mouse model of Alzheimer's disease. *Neurosci. Lett.* **449:** 240–245.
8. Boado, R.J., Y. Zhang, Y. Zhang, *et al.* 2007. Fusion antibody for Alzheimer's disease with bidirectional transport across the blood-brain barrier and abeta fibril disaggregation. *Bioconjugate Chem.* **18:** 447–455.
9. Waxman, S.G. 2010. Polymorphisms in ion channel genes: emerging roles in pain. *Brain* **133:** 2514–2518.
10. Dib-Hajj, S.D., T.R. Cummins, J.A. Black & S.G. Waxman. 2010. Sodium channels in normal and pathological pain. *Ann Rev Neurosci.* **33:** 325–347.
11. Dib-Hajj, S.D., A.M. Rush, T.R. Cummins, *et al.* 2005. Gain-of-function mutation in Nav1.7 in familial erythromelalgia induces bursting of sensory neurons. *Brain* **128:** 1847–1854.
12. Fischer, T.Z., E.S. Gilmore, M. Estacion, *et al.* 2009. A novel Na$_v$1.7 mutation producing carbamazepine-responsive erythromelalgia. *Ann Neurol* **65:** 733–741.
13. Campbell, J.N. & R.A. Meyer. 2006. Mechanisms of neuropathic pain. *Neuron.* **52:** 77–92.
14. Rice, A.S. *et al.* 2008. Animal models and the prediction of efficacy in clinical trials of analgesic drugs: a critical appraisal and call for uniform reporting standards. *Pain* **139:** 243–247.
15. Vierck, C.J., P.T. Hansson & R.P. Yezierski. 2008. Clinical and pre-clinical pain assessment: are we measuring the same thing? *Pain* **135:** 7–10.

16. King, T. *et al.* 2009. Unmasking the tonic-aversive state in neuropathic pain. *Nat Neurosci.* **12:** 1364–1366.

17. King, T., *et al.* 2011. Contribution of afferent pathways to nerve injury-induced spontaneous pain and evoked hypersensitivity. *Pain.*

18. Qu, C. *et al.* 2011. Lesion of the rostral anterior cingulate cortex eliminates the aversiveness of spontaneous neuropathic pain following partial or complete axotomy. *Pain.* in press.

19. Johansen, J.P., H.L. Fields & B.H. Manning. 2001. The affective component of pain in rodents: direct evidence for a contribution of the anterior cingulate cortex. *Proceedings of the National Academy of Sciences of the United States of America* **98:** 8077–8082.

20. LaBuda, C.J. & P.N. Fuchs. 2000. A behavioral test paradigm to measure the aversive quality of inflammatory and neuropathic pain in rats. *Exp Neurol.* **163:** 490–494.

21. Burgess, S.E. *et al.* 2002. Time-dependent descending facilitation from the rostral ventromedial medulla maintains, but does not initiate, neuropathic pain. *J. Neurosci.* **22:** 5129–5136.

22. Devor, M. 2009. Ectopic discharge in Abeta afferents as a source of neuropathic pain. *Exp. Brain Res.* **196:** 115–128.

23. Ringkamp, M. & R.A. Meyer. 2005. Injured versus uninjured afferents: Who is to blame for neuropathic pain? *Anesthesiology* **103:** 221–223.

24. Gracely, R.H., S.A. Lynch & G.J. Bennett. 1992. Painful neuropathy: altered central processing maintained dynamically by peripheral input. *Pain* **51:** 175–194.

25. Devor, M. 1991. Sensory basis of autotomy in rats. *Pain* **45:** 109–110.

26. Rodin, B.E. & L. Kruger. 1984. Deafferentation in animals as a model for the study of pain: an alternative hypothesis. *Brain Res.* **319:** 213–228.

27. Seminowicz, D.A., A.L. Laferriere, M. Millecamps, *et al.* 2009. MRI structural brain changes associated with sensory and emotional function in a rat model of long-term neuropathic pain. *Neuroimage* **47:** 1007–1014.

28. Rodriguez-Raecke, R., A. Niemeier, K. Ihle, *et al.* 2009. Brain Gray Matter Decrease in Chronic Pain Is the Consequence and Not the Cause of Pain. *J. Neurosci.* **29:** 13746–13750.

29. Obermann, M., K. Nebel, C. Schumann, *et al.* 2009. Gray matter changes related to chronic posttraumatic headache. *Neurology* **73:** 978–983.

30. Gwilym, S.E., N. Fillipini, G. Douaud, *et al.* 2010. Thalamic atrophy associated with painful osteoarthritis of the hip is reversible after arthroplasty; a longitudinal voxel-based-morphometric study. *Arthritis Rheum.*

31. Seminowicz, D.A., T.H. Wideman, L. Naso, *et al.* 2011. Effective Treatment of Chronic Low Back Pain in Humans Reverses Abnormal Brain Anatomy and Function. *The Journal of Neuroscience* **31:** 7540–7550.

32. Pereira, F., T. Mitchell & M. Botvinick. 2009. Machine learning classifiers and fMRI: A tutorial overview. *Neuroimage* **45:** S199–S209.

33. Marquand, A., M. Howard, M. Brammer, *et al.* 2010. Quantitative prediction of subjective pain intensity from whole-brain fMRI data using Gaussian processes. *Neuroimage* **49:** 2178–2189.

34. Prato, M., S. Favilla, L. Zanni, *et al.* 2011. A regularization algorithm for decoding perceptual temporal profiles from fMRI data. *Neuroimage* **56:** 258–267.

35. Owen, D.G., Y. Bureau, A.W. Thomas, *et al.* 2008. Quantification of pain-induced changes in cerebral blood flow by perfusion MRI. *Pain* **136:** 85–96.

36. Howard, M.A., K. Krause, N. Khawaja, *et al.* 2011. Beyond patient reported pain: perfusion magnetic resonance imaging demonstrates reproducible cerebral representation of ongoing post-surgical pain. *PLoS One* **6:** e17096.

37. Baron R., A. Binder & G. Wasner. 2010. Neuropathic pain—-diagnosis, pathophysiological mechanisms, and treatment. *Lancet Neurol.* **9:** 807–819

38. Jensen T.S. & R. Baron. 2003. Translation of symptoms and signs into mechanisms in neuropathic pain. *Pain* **102:** 1–8.

39. Maier, C., R. Baron, *et al.* 2010. Quantitative sensory testing in the German Research Network on Neuropathic Pain (DFNS): Somatosensory abnormalities in 1236 patients with different neuropathic pain syndromes. *Pain* **150:** 439–450

40. Baron, R., T.R. Tölle, U. Gockel, *et al.* 2009. A cross sectional cohort survey in 2100 patients with painful diabetic neuropathy and postherpetic neuralgia: Differences in demographic data and sensory symptoms. *Pain* **146:** 34–40

41. Dworkin, R.H., D.C. Turk, S. Peirce-Sandner, *et al.* 2010. Research design considerations for confirmatory chronic pain clinical trials: IMMPACT recommendations. *Pain* **149**(2): 177–193. PubMed PMID: 20207481.

42. Sindrup, S.H., L.F. Gram, K. Brøsen, *et al.* 1990. The selective serotonin reuptake inhibitor paroxetine is effective in the treatment of diabetic neuropathy symptoms. *Pain* **42**(2): 135–144. PubMed PMID: 2147235.

43. Rowbotham, M.C., L. Twilling, P.S. Davies, *et al.* 2003. Oral opioid therapy for chronic peripheral and central neuropathic pain. *N. Engl. J. Med.* **348**(13): 1223–1232. PubMed PMID: 12660386.

44. Irving, G.A., M.M. Backonja, E. Dunteman, *et al.* 2011. A multicenter, randomized, double-blind, controlled study of NGX-4010, a high-concentration capsaicin patch, for the treatment of postherpetic neuralgia. *Pain Med.* **12**(1): 99–109. PubMed PMID: 21087403.

45. Gilron, I., J.M. Bailey, D. Tu, *et al.* 2005. Morphine, gabapentin, or their combination for neuropathic pain. *N. Engl. J. Med.* **352**(13): 1324–1334. PubMed PMID: 15800228.

46. Woolf, C.J. & M.B. Max. 2001. Mechanism-based pain diagnosis: issues for analgesic drug development. *Anesthesiology* **95**(1): 241–249. PubMed PMID: 11465563.

47. Wallace, M.S., M. Rowbotham, G.J. Bennett, *et al.* 2002. A multicenter, double-blind, randomized, placebo-controlled crossover evaluation of a short course of 4030W92 in patients with chronic neuropathic pain. *J. Pain* **3**(3): 227–233. PubMed PMID: 14622777.

48. Dworkin, R.H., D.C. Turk, N.P. Katz, *et al.* 2011. Evidence-based clinical trial design for chronic pain pharmacotherapy: a blueprint for ACTION. *Pain* **152**(3 Suppl): S107–S115. Epub 2010 Dec 9. PubMed PMID: 21145657.

49. Turk, D.C., R.H. Dworkin, R.R. Allen, *et al.* 2003. Core outcome domains for chronic pain clinical trials: IMMPACT recommendations. *Pain* **106**(3): 337–345. PubMed PMID: 14659516.

50. Dworkin, R.H., D.C. Turk, J.T. Farrar, *et al.* 2005. Core outcome measures for chronic pain clinical trials: IMMPACT recommendations. *Pain* **113**(1–2): 9–19. PubMed PMID: 15621359.

51. Senn, S. *Cross-Over Trials in Clinical Research.* Chichester, John Wiley, 1993.

52. Gilron, I. & M.P. Jensen. 2011. Clinical trial methodology of pain treatment studies: selection and measurement of self-report primary outcomes for efficacy. *Reg Anesth Pain Med.* **36**(4): 374–381. PubMed PMID: 21610560.

53. Dworkin, R.H., D.C. Turk, S. Peirce-Sandner, *et al.* 2010. Placebo and treatment group responses in postherpetic neuralgia vs. painful diabetic peripheral neuropathy clinical trials in the REPORT database. *Pain* **150**(1): 12–16. PubMed PMID: 20202753.

54. Rappaport, B.A., I. Cerny & W.R. Sanhai. 2010. ACTION on the prevention of chronic pain after surgery: public-private partnerships, the future of analgesic drug development. *Anesthesiology* **112**(3): 509–10. PubMed PMID: 20124974.